CHEMICAL INDUSTRIES

Series of Reference Books and Text Books

Consulting Editor

HEINZ HEINEMANN

Heinz Heinemann, Inc.,
Berkeley, California

METERING PUMP

METERING PUMPS

Selection and Application

JAMES P. POYNTON

Milton Roy Company
Flow Control Division
Ivyland, Pennsylvania

MARCEL DEKKER   **New York and Basel**

Library of Congress Cataloging in Publication Data

Poynton, James P.
    Metering pumps.

    (Chemical industries ; v. 9)
    Includes index.
    1. Pumping machinery. I. Title. II. Series.
TP159.P8P68 1982   621.6   82-23485
ISBN 0-8247-1759-7

MARCEL DEKKER, INC.
270 Madison Avenue, New York, New York 10016

Current printing (last digit):
10 9 8 7 6 5 4 3 2 1

PRINTED IN THE UNITED STATES OF AMERICA

PREFACE

This volume developed out of a series of articles on metering pumps written during the period 1979-1981. The intent of these articles was to assist engineers involved with precise chemical additive or blending applications to select a cost and performance effective metering pump design.

As revealed in this text, there are many configurations of metering pumps. Major types have evolved to meet specific needs, i.e., hydraulically balanced disc diaphragm liquid ends will operate at high pressure, while mechanically actuated disc diaphragm liquid ends meet low cost requirements. Selection, at best, can be a very confusing process. It is hoped that this volume can provide a ready reference to the advantages and problems associated with a given design, and facilitate preparation of soundly engineered specifications.

Attention is also given to installation and trouble-shooting considerations, with the objective of assisting the engineer who may be experiencing metering pump difficulties in the field. Most start-up problems are due to poor pump suction conditions—the liquid is incapable of keeping up with the plunger on the suction portion of the pump stroke and cavitation occurs. The matter surfaces often in the text.

Chapters 8 and 9 deal with typical metering pump applications. Considerable attention is given to water treatment in all its forms. This discussion is intended to be informative rather than scientific, and to provide the equipment engineer with a ready reference concerning the important role a metering pump plays in various industries.

Credit for this text should go to those who pioneered the development and application of the metering pump since its introduction some 45 years ago.

Instrumental in bringing the book to print was Henri Van Parys of Gray & Rogers, whose material contribution, nudging and encouragement made the task a reality. I also want to thank Milton Roy Company, particularly William S. Ferguson and Jon R. Oliver, for their support. Karl-Eric Strandberg provided crucial graphic and editorial assistance. Finally, I want to thank my fellow workers at all levels in the company for imbuing in me the wherewithal to produce this work.

James P. Poynton

ACKNOWLEDGMENT

Credits not recognized in text are listed below. The extensive contribution of material by Milton Roy Company is particularly appreciated.

The following materials are reprinted from the Milton Roy Company: Figures I.2, 1.3, 1.7, 1.8, 2.1-2.3, 2.6-2.16, 3.1, 3.2, 4.1-4.5, 4.8, 4.9, 5.1, 5.5, 5.7-5.9, 6.1, 6.2, 7.3, 7.5, 7.8, 7.18, 8.2, 8.4-8.6, 9.4, 9.5, 10.1, 10.2; and Table 10.1. The following materials are reprinted from Ref. 8: Figures 1.6 and 2.4; Table 1.1, and Appendix B. The following materials are reprinted from Ref. 1: Figures 1.1, 1.4, 5.2-5.4, and 5.6. The following material is reprinted from Ref. 13: Table 8.1. The following materials are reprinted from Ref. 14: Figures 9.1-9.3. The following material is reprinted from Ref. 3: Figure 1.6. The following materials are reprinted from Ref. 6: Figure 1.5; Table 4.1. The following materials are reprinted from Ref. 7: Figures 4.6, 4.7, and 4.10. The following materials are reprinted from Ref. 5: Figures 1.2, 1.9-1.12, C.1; and Tables 1.2, 3.1, and 3.2. The following material is reprinted from Ref. 4: Figure 3.3. The following materials are reprinted from the Bear Island Paper Company: Figures 7.12 and 7.13. The following material is reprinted from the Foxboro Company: Figure 7.11. The following materials are reprinted from the Pennsylvania Power & Light Co.: Figures 7.1, 7.9, and 7.10. The following materials are reprinted from the Schaefer Brewing Company: Figures 7.14-7.16. The following material is reprinted from the V. F. Weaver Company: Figure 7.17.

CONTENTS

INTRODUCTION

In 1936, Milton Roy Sheen developed the first pump intended to accurately control low-volume delivery of corrosive chemicals against high pressure. This device, called a metering pump, now serves an estimated $175 milion U.S. market and is manufactured by approximately 75 companies. Substantial international demand also exists, particularly in oil-producing and industrialized nations.

The metering pump (also known as the controlled volume or proportioning positive displacement pump) continues to grow in both application and sophistication. Over the years, Sheen's original design has been complemented by myriad configurations and accessories, many of which meet highly specialized requirements.

Virtually every segment of industry employs metering pumps to inject or proportion liquids and slurries. The chemical process industries are by far the largest users; others include electric power generation, petroleum/natural gas, pulp and paper, food and beverage, mining and minerals, and agriculture and horticulture. The pumps also inject water treatment, wastewater treatment and pollution control chemicals, as well as additives to increase the efficiency of fossil fuel combustion.

Chemicals handled by various types of metering pumps range from viscous slurries to solvents, corrosives, oxygen scavengers, radioactive fluids and liquids that must be maintained at elevated temperatures.

I. COMMON METHODS OF METERING

Metering pumps are not the only means available for generating controlled liquid flow rates. However, they often represent the most efficient, most versatile and, in many cases, lowest-cost method. Other devices include positive displacement rotary pumps, loss-in-weight systems, and closed loop centrifugal pump systems.

A. Positive Displacement Rotary Pumps

In positive displacement rotary pumps, the action of the pump forces (rather than induces) the liquid to flow in the required direction. This positive displacement action causes the pump to deliver a consistently similar quantity of liquid with each revolution. Fluid is moved by elements (which may be gears, lobes, vanes, or screws) rotating within an enclosed housing. As the elements rotate, they push the liquid along from the intake of the pump to the point of discharge.

Rotary pumps are especially effective for highly viscous materials, such as molasses or heavy oil, but are not generally used to pump abrasive fluids and are limited in their ability to generate pressures (300–500 psi maximum). They produce a smooth, steady flow and are generally accurate to within ±5% of rated capacity. Variable input from most designs can only be achieved by using an adjustable speed drive.

B. Loss-in-Weight Systems

Loss-in-weight systems belong to the family of gravimetric type feeders that control mass flow rates. A version particularly effective for certain metering applications employs tanks or containers mounted on load cells, with a positive displacement rotary pump used for liquid feed. The latest designs also incorporate a microprocessor to control flow rates, batching quantities and other functions.

Loss-in-weight systems, which yield flow accuracies of ¼ of 1%, are limited to batching operations. Applications mostly involve pharmaceutical and cosmetic production and food processing.

C. Closed-Loop Centrifugal Pump Systems

Centrifugal pumps move fluids by inducing flow. They use a rotating impeller to move liquid from the intake of the pump to the point of discharge by centrifugal force. Approximately 90% of all pumps in use are the centrifugal type because they are the least expensive type of pump. They are commonly used to transfer large volumes of liquid at relatively low pressures where accuracy is not required. Centrifugal pumps handle liquids of low viscosity, deliver a relatively steady discharge flow and are generally accurate to within ±20%.

The method most frequently misused for metering fluids is the closed loop centrifugal pump system. The multi-element closed-loop system is over-qualified for many metering applications because of its high cost, complexity and space requirements.

The simplified closed-loop system shown in Figure I.1 contains a gear pump or centrifugal pump, flowmeter, signal processor, and control valve. As the

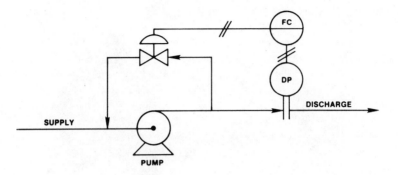

Figure I.1. Multi-element closed-loop system. (Accurate flow metering is achieved by measuring pump output and by-passing excess flow.)

pump operates, flow control is achieved through flowmeter and signal processor interaction that positions the control valve.

The complicated control technique is required because the gear or centrigual pump often provides a higher-than-necessary flow that must be checked at the control valve. The signal processor then determines how much flow the control valve must by-pass to achieve the proper output. A malfunction of any of these components or loss of calibration can cause under- or over-addition of chemicals. This can affect product quality or result in plant shutdown; moreover, over-addition can waste expensive chemicals.

Other closed-loop system drawbacks include high initial and maintenance costs due to the many components involved. Additional restrictions include:

1. Space requirements: Equipment and piping in multi-pump closed-loop systems often requires four-to-eight times as much floor space as a multi-pumphead metering system. Because the equipment is scattered, space is wasted.
2. Continuous process requirements: In multi-element systems, it is difficult to achieve continuous production of metered liquids. Proportioning or flow-rate adjustments must be made constantly, with accompanying stop-start interruptions. Moreover, the practical turn-down ratio or system rangeability is very limited. Once again, the system may require shutdown and component changeout to achieve desired flow rate.

D. Metering Pumps

Metering pumps are another type of positive displacement pump and constitute the theme of this book. Reciprocating metering pumps are accurate to within

Figure I.2   A wide range of standard metering pumps is offered by major manufacturers to meet many requirements. Included are pumps with packed plunger,
disc diaphragm, and tubular diaphragm liquid ends of various sizes that provide
wide capacity ranges and pressure-generating capabilities.

±1% of setpoint and discharge a definite amount of liquid with each stroke of a
piston or plunger. They are used for positive control of liquid flows ranging from
1 ml/hr to 20 gpm and are used against (or to generate) pressures as high as
30,000 psi.

There are two major types of metering pumps. The first, the packed
plunger type, used a direct-acting piston or plunger that moves back and forth in
reciprocating fashion and directly contacts the process fluid within an enclosed
chamber. The second, the diaphragm type, uses the reciprocating motion of the
piston or plunger to transmit pressure through hydraulic fluid to a flexible diaphragm. The diaphragm isolates and displaces the pumped fluid.

This book is devoted exclusively to reciprocating metering pumps and
chemical feed systems that employ versions of these mechanisms, and is

intended to clear up much of the confusion surrounding a rather complex class of pump products.

Modern industry demands process efficiency—the information contained herein can help the engineer attain that efficiency with respect to the metering function. Answers to questions such as "Which configuration is best for me?" and "Where should I use a metering pump?" are provided on an industry-by-industry basis. Special considerations are covered, and maintenance guidelines are provided. Finally, for anyone with the energy bee in his bonnet, several suggestions for reducing pump electrical consumption are detailed.

METERING PUMPS

1

PRINCIPLE OF OPERATION

The terms controlled volume pump, metering pump and proportioning pump are often interchanged when referring to a specific family of devices within the broad category of positive displacement pumps. For simplicity, the term metering pump will serve as the general nomenclature for any chemical pump offering inherent steady state delivery accuracy, pressure generating capability and output that is adjustable while in operation.

Metering pumps are subdivided into various classifications that depend primarily upon the configuration of the displacement chamber or liquid end. A list of the more basic liquid end types follows:

1. Packed plunger.
2. Diaphragm
   - Disc diaphragm
   - Tubular diaphragm
   - Double diaphragm
     - Double tube
     - Disc/tube
     - Double disc.

The packed plunger liquid end contains a piston that directly contacts the pumped fluid. The diaphragm or bellows type liquid end features some type of positive barrier arrangement that prevents the pumped fluid from contacting the plunger. Both configurations include up to three additional mechanisms:

1. A prime mover, either electric or pneumatic.
2. A drive mechanism with built in manual capacity control.
3. A device for automatic capacity control (usually optional).

Pumps may be driven by an electric motor, the most common drive source, or a pneumatic cylinder with the cylinder rod directly coupled to the plunger. Pneumatically driven pumps are particularly useful in hazardous locations, areas in which electric power cannot be utilized and where digital control techniques are desirable for controlling pump delivery.

Driver mechanisms include variable slider crank, as well as those employing mechanical and hydraulic lost motion principles. Capacity adjustment is achieved via a manual micrometer control knob or by pneumatic or electric controllers in response to process feedback. Capacity may be further refined through the use of variable speed motors.

I. THE PRIME MOVER

The constant-speed AC electric motor is the most commonly used driver in metering pumps, with the type of motor normally specified by the purchaser and sized by the supplier to meet maximum specified operating conditions. Motors are either flange-mounted and coupled to the pump input shaft or foot-mounted and supported on a bracket attached to the pump casing.

DC electric motors are mounted similarly (although greater weight per unit hp sometimes limits flange-mounted configurations) and are usually used in conjunction with motor controllers. Silicon-controlled rectifier (SCR) circuits in the controllers vary motor speed and, hence, pump output, and can be paced automatically from a remote analog signal.

Pneumatic power is sometimes used to drive either a plunger or diaphragm liquid end. This type of drive uses a reciprocating pneumatic cylinder sized to provide the necessary thrust differential between driving force and generated pressure. Pressure to the cylinder piston is most often provided by a four-way valve that alternately admits pressure to opposite sides of the cylinder piston. Actuation of the four-way valve may be accomplished by electrically energizing a solenoid, which changes direction of the pressure porting, or through an internal pressure porting arrangement to automatically cycle the cylinder piston. Pneumatic power facilitates use of digital control techniques because electrical on-off pulses to the solenoid are directly analogous to plunger reciprocation.

Metering pumps with solenoid drives are available for those applications requiring low horsepower. Although such pumps offer high turn-down ratios and interface directly with control signals, limited solenoid cycle life constrains use of this pump to noncritical applications.

II. THE DRIVE MECHANISM

The capacity of a metering pump is a function of the diameter of the plunger, the effective length of the plunger stroke and the rate or speed of stroking. Since the plunger diameter is constant for a prescribed pump, stroking length and pump

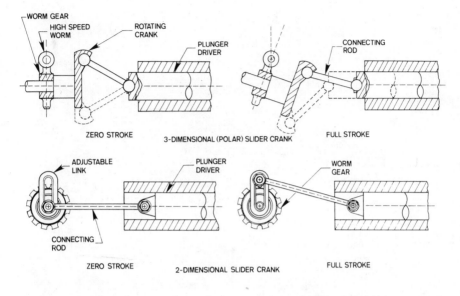

Figure 1.1. Examples of slider crank mechanisms that can be adjusted to alter the volume of displaced fluid.

speed are the variables available to adjust the output capacity while the pump is operating. Speed and/or stroke length adjustments can be effected manually or automatically, in accordance with the process demand, to vary the volume of liquid delivered.

Reciprocating pumps are so named because their operation depends upon the back-and-forth action of a plunger, resulting in displacement of process fluid directly or as a result of diaphragm flexure. A principal characteristic of metering pumps is adjustable output. This is made possible by varying the actual or effective length of the displacement stroke.

There are two main classifications for stroke length adjustments. One involves varying the radius of eccentricity of the plunger drive mechanism, which is sometimes referred to as amplitude modulation. This method includes various 2 and 3 dimensional slider crank configurations (see Figure 1.1) and the shift ring drive. The second method involves the portion of fixed crank travel which is transmitted for effective plunger displacement. Included here are mechanical lost motion with eccentric cam and hydraulic by-pass control configurations. Distinct advantages and disadvantages are inherent in each design (see Figures 1.2, 1.3, and 1.4).

The instantaneous flow characteristic from a reciprocating plunger mechanism approximates a sine wave, as shown in Figure 1.5. Adjustable crank radius drives reduce sine wave amplitude and associated inertial and frictional effects at

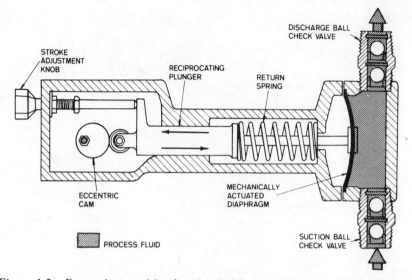

Figure 1.2. Eccentric cam drive (mechanical lost motion) provides for changes in the discharge flowrate through variation of the plunger return position. Crank eccentricity remains constant through the entire flow rate.

reduced capacity settings. Lost motion drive performance is characterized by sharp flow transitions at reduced capacity settings. In lost motion drives, inertial effects are significant at all capacity settings, while frictional effects are maximum and unchanged at settings above 50%.

A. Amplitude Modulation

The slider crank mechanism, a forerunner of most modern drive designs, features a turnbuckle-type adjustment that permits the stroke length to be altered by changing length of a pivot arm. Many variations of this drive have evolved. Although they are associated with different names and trademarks, the same basic operating principle is readily recognizable. All slider crank mechanisms are attached to the plunger, though the various methods of linkage may differ. This drive is generally utilized in the higher-capacity pumps that operate under greater pressures.

    A unique shift ring drive is offered by one manufacturer to minimize mechanical vibration and shock loads in the drive train. In this design, a spring-loaded plunger rides within the main drive shaft, and reciprocation is caused by adjusting the position of a ring within which the plunger rotates.

    For less demanding applications, an alternate method was developed to transfer power to the pumping element with equal accuracy. This method, employing the lost motion principle, is described in the following section.

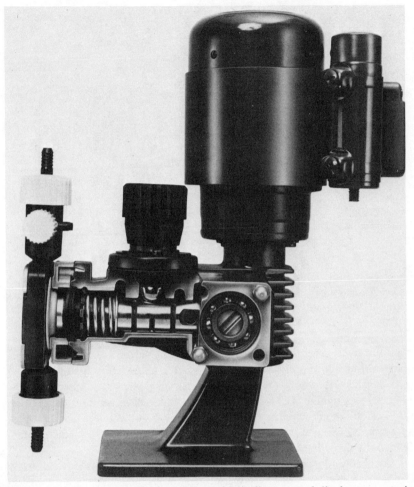

Figure 1.3. Cutaway view of typical mechanically actuated diaphragm metering pump.

B. Lost Motion Drives

As the name implies, lost motion drives provide a method of altering the stroke length by somehow "losing" or not utilizing the drive thrust available over a full pumping cycle. Lost motion drives are further subdivided into hydraulic and mechanical types.

Hydraulic lost motion entails a change in the effective rather than actual stroke length. The plunger continues to reciprocate the full length of the stroke

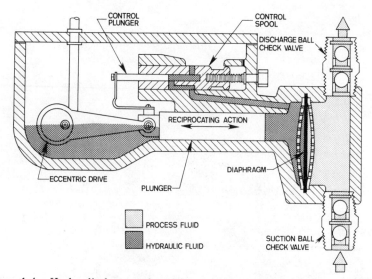

Figure 1.4. Hydraulic lost motion drive permits a portion of hydraulic fluid to escape through a by-pass valve with each stroke, permitting the effective stroke length of the plunger to be changed while the actual stroke length remains constant.

at all times, but some of the hydraulic fluid available to deflect the diaphragm escapes through a bypass valve that returns this unused fluid to the sump or hydraulic fluid reservoir. This bypass valve can be externally adjusted while the pump is in operation. The lost motion drive mechanism can be built at lower cost than the more complex slider crank, but it is not intended for use where capacities in excess of 150 gph are encountered because of hydraulic hammer effects at by-pass valve cut-off.

Mechanical lost motion differs in principle—the diaphragm is mechanically attached to the plunger or deflecting mechanism. The mechanism is driven forward by an offset cam, causing the diaphragm to deflect a predetermined amount. The flow is varied by limiting the return of the mechanism. Generally, a spring causes the driving mechanism to return to its original position after each cycle of the offset cam. If the plunger is allowed to return only half-way to the original position, the stroke and accompanying deflection will be decreased by 50%.

Mechanical lost motion is not generally used when pressures exceed 250 psi or when capacity requirements exceed 75 gph. These pressure and capacity restrictions result because the diaphragm is not hydraulically balanced; however, pumps employing this type of actuation tend to be less costly than their hydraulically balanced counterparts.

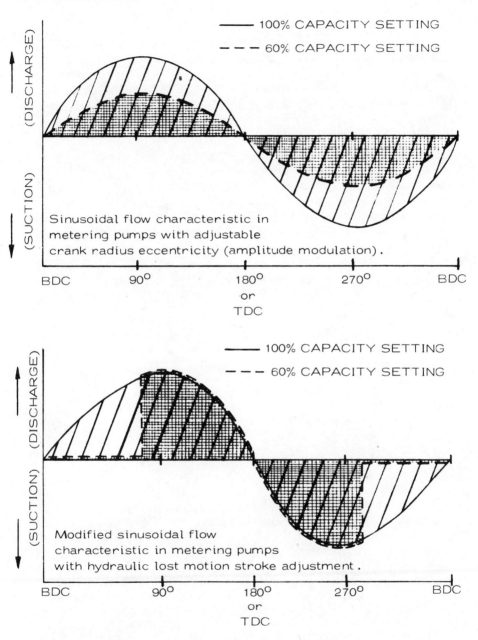

Figure 1.5. Metering pump flow characteristic. Fluid displacement and flow velocity vary sinusoidally with maximum rates occurring approximately half way thru suction and discharge cycles.

III. PACKED PLUNGER PUMPS

The packed plunger liquid end is the basic design from which all types of liquid ends originate. Pumping is accomplished by one or more sealed plungers that draw in and expel fluid. One-way check valves on the inlet and outlet ends of pump operate 180° out-of-phase to control filling of the displacement chamber during the suction or "vacuum" stroke and to prevent backflow to the supply system during the discharge stroke (see Figure 1.6).

As the plunger reciprocates, the pumped liquid is alternately drawn into and discharged from the liquid end. Each suction (rearward) stroke of the pump plunger creates a negative pressure in the displacement chamber. The pressure of the liquid in the suction line unseats the suction ball checks and liquid flows into the displacement chamber. On the discharge stroke, the plunger moves forward and pressurizes the liquid, which unseats the discharge ball checks and permits flow out of the discharge port. On each suction stroke, the discharge ball checks are seated, and on each discharge stroke, the suction ball checks are seated (pressure in pump head is greater than suction line pressure). This mode of operation prevents backflow and ensures liquid movement from suction port, through the displacement chamber, and out of the discharge port.

The packed plunger design has several advantages over other configurations. Among them are low initial cost, simplicity of design and the capability to

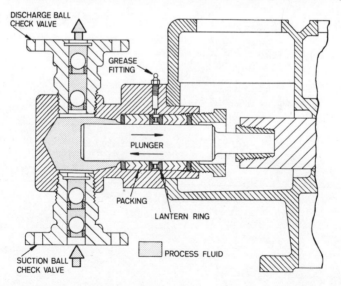

Figure 1.6. Packed plunger liquid end features simple, high-performance design but requires periodic maintenance and allowance for a small amount of leakage to cool the packing.

develop high discharge pressures. In addition, the ability of the sealed piston to lift fluids to the pump inlet is limited ultimately only by the vapor pressure of the fluid. The plunger pump, in its basic form, may be operated at pressures to 30,000 psi, which repetitive accuracy to within ±1% of rated capacity.

This design has several inherent disadvantages, however, which make it a poor choice for certain applications. One drawback is in the design of the plunger mechanism, which requires packing to effect a tight seal between the plunger and plunger bore. A small amount of controlled leakage of process fluids must be tolerated in most instances to cool and lubricate the plunger. Additionally, the unavoidable friction that results as the plunger reciprocates wears the packing and increases the leakage of process fluid past the plunger. Periodic adjustment is necessary, since such wear decreases the volumetric efficiency of the pump. Replacement will eventually be necessary as the packing progressively deteriorates, further contributing to maintenance costs.

The packing material used depends on a number of factors, including temperature, pressure, liquid composition and pump speed (see Table 1.1). If abrasive slurries are to be pumped, lantern rings should be provided as a means of flushing the packing. Chevron, lip seal, and square ring packing are common types used with plunger pumps, with a wide variety of materials available in each configuration.

In summary, packed plunger pumps are employed in those applications requiring a low cost pump, and where a certain amount of leakage and maintenance is permissible. Applications characteristically involve injection of a nontoxic and non-corrosive chemical in both batching and continuous operations. However, the increasing need for a method of moving corrosive acids, as well as federal regulations and environmental concern, has forced pump manufacturers to develop the sealed diaphragm configuration that, although more costly to produce, overcomes many of the disadvantages of the packed plunger design.

IV. DIAPHRAGM PUMPS

Diaphragm metering pumps provide an effective solution to the leakage problem associated with packed plunger pumps. The diaphragm acts as an interface between the plunger and the process fluid. The plunger still travels through a bore in this design, but instead of directly displacing the process fluid, it is used to actuate a diaphragm across which energy is transmitted. On the opposite side of the diaphragm, one-way check valves permit a proportional amount of process fluid to be drawn into the displacement chamber and discharged with each cycle of the plunger.

Diaphragms are either mechanically actuated (i.e., directly coupled to the plunger) or hydraulically actuated. The hydraulically actuated diaphragm is balanced between two fluids, thereby eliminating diaphragm fatigue problems and

TABLE 1.1

Typical Packings for Various Process Fluids

Fluid to be Pumped	Maximum Temperature	Maximum Pressure	Typical Packing Type	Description
All liquids	500°F	2000 psi	Automatic	Solid-ring chevron; molded pure Teflon*
Aqueous solutions; not for aromatics or strong aqueous solutions of acids or bases.	200°F	2000 psi	Automatic	Solid-ring chevron; neoprene duck
High purity water; not for aromatics or strong aqueous solutions of acids or bases.	300°F	2000 psi	Automatic	Split-ring chevron; neo-asbestos
High purity water; strong acid solutions; some solvents.	500°F	4000 psi	Compression	Square-cut braided; impregnated asbestos-Teflon*
High purity water; special liquids.	300°F	5000 psi	Semi-automatic	Split-cup asbestos-Teflon* duck
Food products; beer.	250°F	2000 psi	Automatic	White rubber duck
Mineral base oils; gasoline, kerosene, fuel oil; water and water-emulsion hydraulic fluids.	250°F	7500 psi	Automatic	Solid-ring "Vee"; Buna N and cotton duck
Strong acids; not for solvents.	300°F	250 psi	Compression	Square-cut braided; blue asbestos
High-pressure applications; corrosive acids and alkalis; coal slurries.	500°F	5000 psi	Automatic	Solid-ring chevron; molded Teflon*

TABLE 1.1 *(continued)*

Fluid To Be Pumped	Maximum Temperature	Maximum Pressure	Typical Packing Type	Description
Strong corrosives; heat transfer fluids; water	1500°F	4000 psi	Compression	Split-ring; graphite

Packings shown are typical. Specific recommendations are based on actual application. For example:

1. Use close-clearance metal adapters and back-up rings for pressures of 1000 psi and higher.
2. Use close-clearance rings and hard plungers for pressures of 2000 psi and higher.
3. Whenever flushing is required, a chevron packing must be used.
4. For lime and diatomaceous earth slurries, use neoprene duck and flush connections.
5. Pumps metering demineralized or deionized water cannot be lubricated, since any greases added to the packing may contaminate the pure water. Consequently, pumps for this service must be equipped with ceramic plungers and Teflon* QP packing.

*Registered DuPont trademark

permitting higher discharge pressures.

In order for the hydraulically actuated liquid end to maintain accuracy, the proper volume of hydraulic fluid must be kept in the hydraulic chamber. A three-valve system is often used to ensure that the hydraulic fluid is relatively free of air or entrained gases and to protect against overpressure in both hydraulic and process fluids:

1. A refill valve, which also acts as a vacuum breaker, is used to maintain constant volume of hydraulic fluid by replenishing any that leaks out of the chamber.
2. A gas bleed valve removes air or gas bubbles from the hydraulic fluid.
3. A relief valve in the hydraulic chamber provides overpressure protection for both hydraulic and process fluids.

Controlled porting of hydraulic fluid may also be used to accomplish refill and gas bleed action.

Check valves located on the suction and discharge sides of the pump admit and expel the process fluid in response to the negative and positive pressure exerted by the flexing diaphragm. Double check valves are often utilized to increase the accuracy of the pump and to provide redundant sealing action.

A. Disc Diaphragm Liquid Ends

The hydraulically actuated disc diaphragm is generally made of TFE (tetrafluoro-ethylene) and flexes between two dish-shaped supporting (contour) plates with flow-thru holes. The contour plates provide diaphragm containment and prevent it from rupturing under high pressure.

Diaphragm metering pumps are mainly used to handle liquids in applications where even minimal leakage is not acceptable. These pumps often employ the hydraulic lost motion principle. Pumping action is developed and controlled by four basic components (see Figure 1.7):

1. The pump plunger "A," which reciprocates with a constant stroke length and displaces oil into and out of the diaphragm chamber "C."
2. The flexible diaphragm "X," which is a movable partition between the plunger oil and fluid being pumped.
3. An oil by-pass circuit from the diaphragm chamber "C" to the reservoir "D" through passage "E," by-pass port "H" and control valve "F."
4. A by-pass control plunger "G," which moves with and is directly coupled to the pump plunger to correlate by-pass shut-off at port "H" to pump plunger position.

In operation, as the pump plunger and by-pass control plunger move forward as shown in Figure 1.7, the displaced oil is by-passed to the oil reservoir until the control plunger "G" closes the by-pass port "H" as shown in Figure 1.8. Then

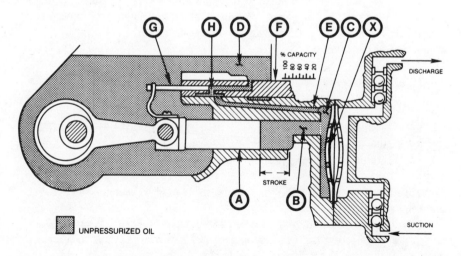

Figure 1.7. Disc diaphragm liquid end.

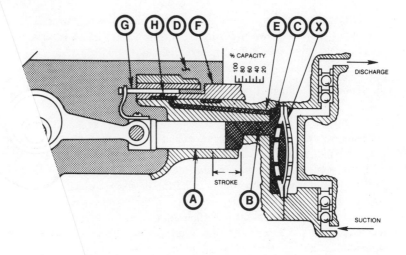

F   lost motion design showing fixed stroke piston, by-pass
po

the   er displacement is imposed on the flexible diaphragm
which   s the fluid being pumped through the discharge ball
checks

O,   ·, the pump plunger pulls oil out of the diaphragm
cavity w   ible diaphragm and pulls fluid through the suction
ball check   plunger "G" opens the by-pass port "H," the bal-
ance of the   ·ment can be supplied from the reservoir through
the by-pass p ıssa₋·.

The discharge capa   ljusted from 0-100% by rotating the adjustment
knob, which moves the contro ılve "F" so that the by-pass port "H" is closed
at the desired percentage of the total plunger stroke. When the control valve is
adjusted for 50%, the by-pass port will be positioned so that it is opened when
the plunger has completed one-half of the suction stroke. On the next pressure
stroke the oil displaced by the pump plunger will be by-passed through the open
port to the reservoir for the first 50% of the stroke before the by-pass port is
closed by the control plunger. The remaining 50% of the plunger displacement is
imposed on the flexible diaphragm so that fluid is discharged for only 50% of
the plunger travel. A similar analysis would apply for the 0% capacity setting on
the control valve where all the plunger oil displacement is by-passed to the
reservoir.

Mechanically actuated disc diaphragm liquid ends are also available but are
somewhat limited in terms of pressure and capacity (see Section II-B, Lost

Motion Drives). In this configuration, the diaphragm is mechanically coupled to the pump drive as shown in Figure 1.2. Figure 1.3 is a cutaway view of a typical mechanically actuated diaphragm metering pump.

B. Tubular Diaphragm Liquid Ends

In the tubular diaphragm design, a plunger reciprocates as described above, but a tube-shaped elastomeric diaphragm expands or contracts with the pressure provided by the hydraulic fluid. Contraction or expansion of the tube, combined with one-way action of suction and discharge check valves, creates a pulse of metered liquid through the pump. Cavitation problems are reduced since no contour plate is used on the process side of the tube.

In the tubular diaphragm configuration shown in Figure 1.9, the plunger is located within the elastomeric tube itself. During the suction stroke, a vacuum is created within the tube when the plunger is withdrawn. This causes the tube to contract, drawing a predictable amount of process fluid through the suction check valve and into the displacement chamber located about the exterior of the diaphragm. The opposite occurs during the discharge stroke—pressurization of the hydraulic fluid within the tube causes it to expand and displace a proportional amount of fluid through the discharge valve.

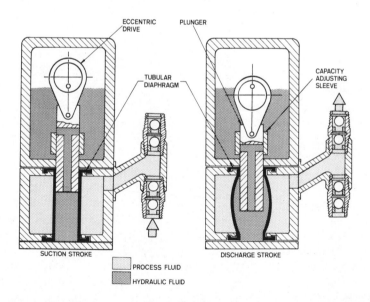

Figure 1.9. Single tubular diaphragm liquid end. This design, employing the hydraulic lost motion drive concept, features excellent suction lift capability.

1) Tubular Diaphragm Variations

Variations include the double tube (Figure 1.10) and combination disc/tube (Figure 1.11). In these twin-diaphragm pumps, the plunger imparts hydraulic pressure to flex a primary diaphragm which, in turn, flexes a secondary diaphragm via an intermediate liquid. The secondary diaphragm (tube) handles the process fluid while isolating it entirely from the pump head, precluding the need for costly liquid end materials to resist corrosion by the process fluid. Further, the secondary diaphragm may be considered an extension of the suction and discharge piping, providing little resistance to flow. Such a pump is ideally suited for metering viscous liquids and abrasive slurries and is not prone to sludge accumulation.

Twin-diaphragm pumps offer the added safety of a back-up diaphragm should the process diaphragm fail; an inert hydraulic fluid is used between the diaphragms. This latter feature is useful when handling products that may react violently with hydraulic oils or when contamination of the process chemical must be prevented in the event of secondary diaphragm failure.

The intermediate chamber between the two diaphragms is sometimes equipped with a sight glass to allow monitoring. If the chamber is filled with a liquid of known pH reference color, the change of color warns of diaphragm failure. Electrodes may also be provided in the intermediate chamber to monitor

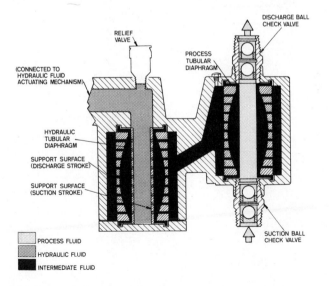

Figure 1.10. Double tubular diaphragm liquid end. This design features flow-thru construction, while minimizing surface area of wetted metallic parts.

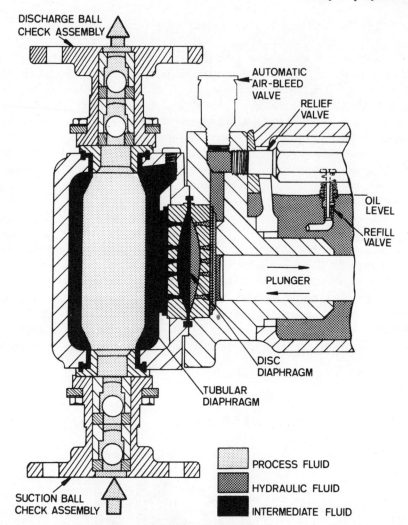

Figure 1.11. Disc/tubular diaphragm liquid end. This design offers features similar to the double tubular diaphragm design.

conductivity of the intermediate fluid. A change in conductivity can be detected by the electrodes to provide a warning of diaphragm failure.

C. Miscellaneous Configurations

Other liquid end configurations include the double TFE disc diaphragm (see Figure 1.12). This design with the inherent extra safety twin diaphragms offer,

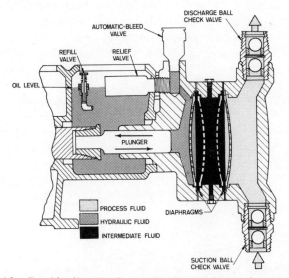

Figure 1.12. Double disc diaphragm liquid end design features safety inherent in redundant diaphragms, universal compatibility of TFE diaphragm material, and absence of process contour plates.

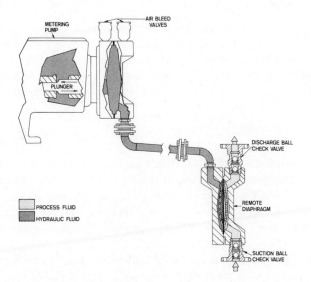

Figure 1.13. Remote diaphragms permit hazardous products to be handled safely.

is particularly well suited for handling solvents and other related chemicals that attack elastomeric tube materials. The primary disadvantage of such a design lies in the requirement for potentially expensive metallurgical construction.

Both double and single diaphragm heads may be remotely located to handle hazardous products such as high temperature, cryogenic and radioactive fluids (Figure 1.13). Remote assemblies may also be employed where space does not permit installation of a complete unit, or to lift corrosive fluids from top-opening tanks.

All of these diaphragm designs are in use in various processes throughout industry. Selection of the most suitable design for any application depends on the characteristics of the fluid being handled, the amount of maintenance permissible, safety aspects and desired system cost (see Chapter 3, Selecting Metering Pumps).

V. OTHER OPTIONS

The options previously discussed are complemented by a host of metallurgical choices for wetted parts. Liquid ends may be fabricated from plastics, ferrous, alloy or superalloy materials. Available materials of construction for the various types of liquid ends are listed in Table 1.2. Together with the preceding liquid end and drive configurations, more than 5,000 permutations can be developed from two basic types of reciprocating metering pumps—packed plunger and diaphragm.

Other options include pulsation dampeners, back pressure valves, relief valves, automatic valves and controls, metering rate timers, tanks—the list goes on and on. These options, which are discussed in Chapter 2, are often combined with metering pumps to form prepackaged systems that are custom-engineered to perform a specific function. Chapter 4 is devoted to these systems.

VI. TOWARD AUTOMATION

Due to the overwhelming number of applications and permutations, some metering pump manufacturers are using data processing (DP) to facilitate specification. Programs have been developed to calculate net positive suction head (NPSH) based on parameters supplied by the user (see Chapter 6, Installation). The computer, in effect, verifies that the pump will perform properly when it is installed in the field.

Computerized testing has also been introduced as an effective means for verifying accuracy of each metering pump prior to shipment. The system uses a computer to monitor the flow of each pump during several test runs at operating pressure to determine repetitive accuracy. When a pump "tests out" (i.e., falls within acceptable parameters), a printer supplies hardcopy listing all performance data. This printout is then given to the pump user.

TABLE 1.2

Available Materials of Construction*

Pump Type	Liquid End	Diaphragm
Mechanically Actuated Disc Diaphragm	H through I	M–P
Hydraulically Actuated Disc Diaphragm	A through L	N
Tubular Diaphragm	A through K	M, O, P
Tubular/ Disc Diaphragm	A through J	M, O, P
Double Tubular Diaphragm	A through K	M, O, P
Double Disc Diaphragm	A through K	M, O, P

Material List

A. Carbon Steel
B. Grey Iron
C. Ductile Iron
D. 304 SS
E. 316 SS
F. Type 20 SS
G. Nickel-based Alloys A, B and C
H. PVC
I. Polyethylene
J. Polypropylene
K. Acrylic
L. Polyvinylidene fluoride
M. FLU (fluorinated hydrocarbon)– Viton**
N. TFE (tetrafluoroethylene)– Teflon**
O. EPDM (ethylene propylene terpolymer)
P. CSM (chlorosulfonated polyethylene–Hypalon**
Q. Ceramic

*Individual manufacturers may offer specialized designs in which other materials are employed.
**Registered DuPont trademark.
Source: Milton Roy Company.

2

METERING PUMP ACCESSORIES

A wide variety of accessories may be specified for use with metering pumps and chemical feed systems to improve flow characteristics, provide added system protection and/or allow difficult pumpables to be accurately metered.

I. ACCESSORIES

A. Back Pressure Valves

A typical back pressure valve is shown in Figure 2.1. Metering pumps that discharge into open tanks or other systems in which discharge back pressure is less than 30-50 psig should be equipped with back pressure valves to ensure repetitive metering accuracy. In some manufacturers' pumps, an artificial discharge head pressure can be created through installation of a back pressure spring in the discharge check valve; in others, installation of a back pressure valve in the discharge line is necessary. Normally, the valve is located near the pump; however, back pressure valves for large pumps with long and extremely small discharge lines may have to be installed near the point of discharge into the process. Individual manufacturers should be consulted to determine the preferred method of installation.

B. Pulsation Dampeners

When back pressure valves are used to provide an artificial discharge head, a pulsation dampener should be included in the discharge line to smooth out flow from the pump to the back pressure valve by relieving flow peaks. This extends valve life and provides a more uniform flow rate of process liquid.

Typical pulsation dampeners are shown in Figures 2.2 and 2.3. Without the pulsation dampener, the valve mechanism will snap open and closed with the surge from each pump stroke. The pulsation dampener allows the back pressure valve to oscillate about a partly closed position, thus minimizing wear.

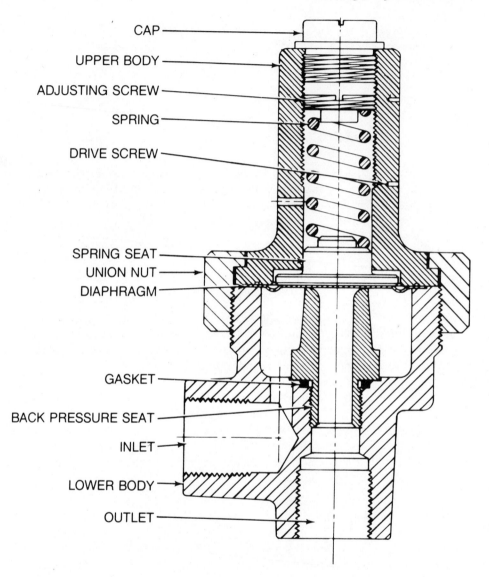

Figure 2.1.   Typical back pressure valve.

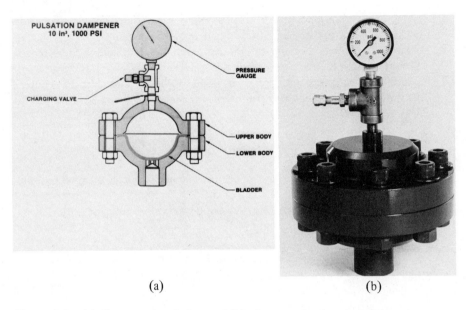

(a)   (b)

Figure 2.2. (a) Cross sectional view and (b) photograph of typical high-pressure pulsation dampeners (up to 1,000 psig).

Figure 2.3. Typical low-pressure pulsation dampeners (up to 250 psig).

Discharge line pulsation dampeners offer the additional advantage of limiting the flow and pressure variations inherent in reciprocating metering pumps. Reciprocating metering pumps exhibit a sinusoidal, pulsating flow characteristic that represents what the plunger is doing—speeding up, slowing down or reversing direction. Although lovers of kinematics may be entranced by the complex differential equations that describe this phenomenon, the practical consequences can be troublesome:

1. Pulsating flow can cause pipes to rattle and hammer, which leads to structural damage or leakage.
2. Premature pump breakdown can occur due to cyclic loading on bearings, valves and other critical components.
3. Viscous and acceleration losses can be severe, resulting in cavitation and a need for oversized piping.

Installation of a properly sized pulsation dampener can eliminate these problems and may reduce system costs dramatically by permitting substitution of smaller diameter discharge piping.

C. Vented Risers

Most reciprocating pumps cannot deliver controlled flow unless the discharge line pressure is at least 15 psi greater than the suction line pressure. If necessary, artificial discharge pressure can be created by installing a pulsation dampener and back pressure valve as previously described.

In many applications, a vented riser can serve the same purpose at a much lower cost. A vented riser (Figure 2.4) is simply a vertical extension of the discharge pipe into an open tee. The other side of the tee goes to process. Practically maintenance-free, this device prevents siphoning and provides an artificial discharge head; however, a clogged or closed line may cause the riser to overflow. Therefore, the pulsation dampener/back pressure valve assembly should be substituted when hazardous liquids will be handled.

D. Multiplexing

Figure 2.5 is a graphic representation of the flow pattern that is characteristic of simplex, duplex, triplex, quadruplex and quintuplex liquid end arrangements. These arrangements have one, two, three, four and five plungers, respectively. (In multiplexed configurations, a single driver powers two or more plungers that are phased to minimize flow peaks.) Note that flow surges are most severe in simplex and duplex versions, diminish dramatically in a triplex, worsen a bit in a

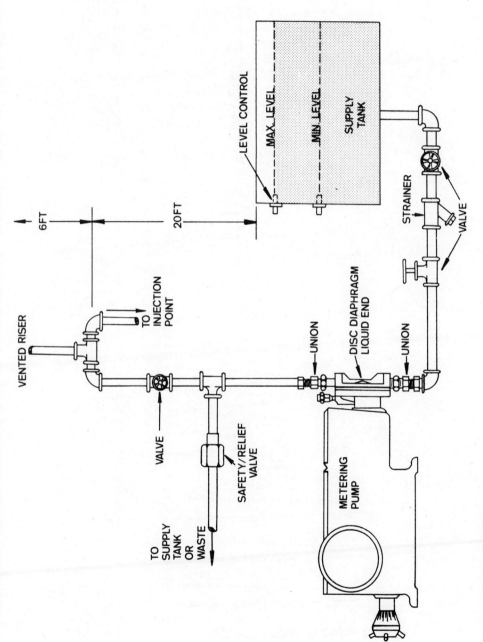

Figure 2.4. Vented riser.

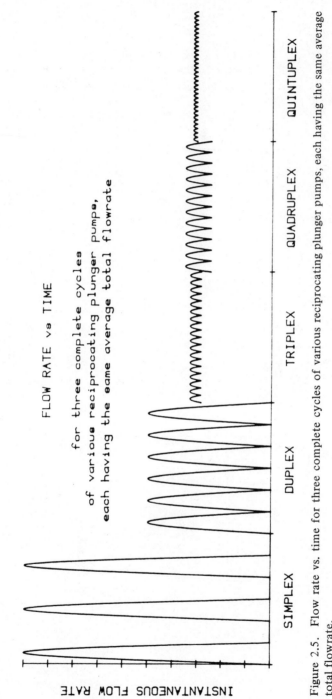

Figure 2.5. Flow rate vs. time for three complete cycles of various reciprocating plunger pumps, each having the same average total flowrate.

quadruplex and really smooth out in a quintuplex. In lieu of a pulsation dampener, the use of a pump with three or more properly phased plungers can reduce
pulsation to a generally acceptable level. Multiplexing also offers the opportunity for energy savings (see Chapter 9, Energy Consideration).

E. Safety Valves

Motor driven metering pumps can build up high discharge pressures in one or
two strokes, and thermal overload or similar safety devices in the motor starter
circuit often act too late to prevent damage to the pump, discharge line or
process equipment when the discharge line becomes obstructed. To prevent such
occurrence, the discharge line should be equipped with a suitable safety valve or
rupture device sized to handle system pressures and to discharge the maximum
pump flow rate safely. A typical safety valve is shown in Figure 2.6.

The safety valve should be located in the discharge line between the pump
and the first downstream shut-off or back pressure valve (this prevents pump
damage from accidental valve closure). Outlet of the safety valve is then piped
back to the suction supply tank or to a suitable disposal point. The open end of
the return piping should be located where it is visible so any safety valve leakage
can be detected.

F. Check Valves

Metering pumps are normally supplied with one-way suction and discharge check
valves (Figure 2.7) that operate 180° out-of-phase. The suction check valve permits fluid to be drawn in during the suction stroke of the plunger and prevents
backflow during the discharge stroke; the discharge check valve permits fluid to
be expelled during the discharge stroke and seats during the suction stroke. Installation of an addition check valve where the discharge line enters a boiler or
similar high-pressure vessel is recommended. This prevents backflow through the
discharge piping and isolates the pump discharge from system pressures (a safety
consideration).

G. Remote Diaphragm Assemblies

The liquid end assembly, including the diaphragm head(s), contour plates and
diaphragm(s) can be remotely located to improve net positive suction head conditions or for handling low-temperature fluids such as liquified gases (Figure
2.13). These remote assemblies may incorporate disc, tubular or double diaphragm assemblies, depending on the nature of the process fluid.

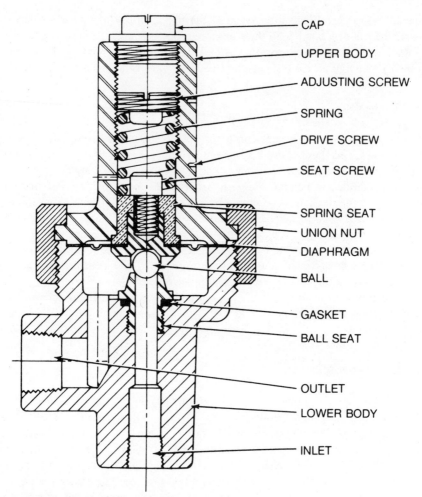

Figure 2.6. Typical safety valve.

H. Sludge Traps

When concentrated sulfuric acid is metered, sludge particles tend to accumulate on the pump check valves and prevent them from performing properly. Sludge traps have been specifically designed to eliminate this problem at flows to 20 gph. A typical sludge trap is shown in Figure 2.8. This trap is designed with three filtering areas—a settling chamber, an intermediate prefilter of ceramic balls and a final fiberglass filter. This design prevents the fiberglass final filter from becoming rapidly clogged with larger sludge particles, since most of the sludge will settle to the bottom of the trap or be held back by the ceramic ball prefilter.

Figure 2.7. Plastic check valve disassembled (top) and installed on metering pump (bottom).

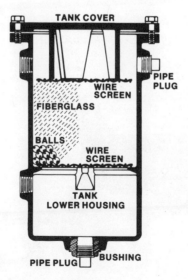

Figure 2.8. Typical sludge trap contains three filtering areas—settling chamber, intermediate prefilter, and final filter.

The velocity of the concentrated sulfuric acid entering the sludge trap inlet port is reduced in the settling chamber. Because of the reduced flow rate, the largest sludge particles cannot remain suspended and will settle to the bottom of the trap. Particles which remain suspended are, for the most part, retained in the trap when the acid flows upward through several layers of 3/8-in. ceramic balls supported by a heavy wire screen. Only very fine sludge particles remain in the acid, and most of these are trapped in the thick fiberglass final filter. After passing through the final filter element, the virtually sludge-free acid passes through the outlet port to the pump.

For minimal maintenance and ease of cleaning without interruption of acid flow to the pump, the sludge trap should be installed with a simple manifold piping arrangement that allows for isolation and periodic back flushing. Such an arrangement is shown in Figure 2.9.

I. Leak Detectors

Hazardous or explosive fluids often dictate use of a twin-diaphragm pump equipped with a device for alerting the user to diaphragm malfunction. In these applications, a conductivity sensor type leak detector is mounted in the intermediate chamber between the primary and secondary diaphragms. The leak detector circuit shown in Figure 2.10 includes a probe or sensor, control relay and annunciator/pump control. The probe responds to increases or decreases in

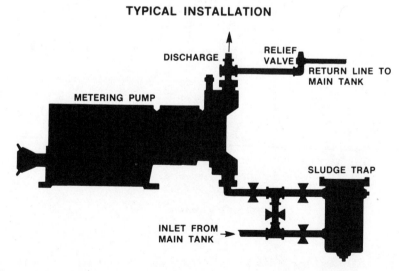

Figure 2.9. Typical sludge trap installation.

TABLE 2.1

Leak Detector Application Examples

Specific Resistance of Process Fluid in Ohm Centimeters	Typical Process Fluid	Recommended Intermediate Liquid
1,000	Concentrated and dilute acids and alkalis, cleaning and plating solutions, sea water, ammonia, ammonium chloride.	
2,200	Bear and wine.	75% Glycol
4,700	Hard water from wells.	25% Distilled water
10,000	Lake and river water, treated municipal water, sewage.	
22,000	Soft water from wells, process steam condensation.	

Note. This system is not recommended for process liquids above 22,000 ohm centimeters specific resistance. (For example: corn syrup − 47,000 ohm − cm, distilled water − 47,000 ohm − cm, demineralized water − 1,000,000 ohm − cm.)

Source: Milton Roy Company

intermediate liquid conductivity caused by process fluid contamination. The sensor may be connected to the user's own system or to a conductivity actuated ON-OFF relay to halt the entire process, sound an alarm or trigger some other alert mechanism. Process fluids that dictate use of a leak detector are listed in Table 2.1.

As previously mentioned, the intermediate chamber can also be equipped with a sight glass to allow the pH reference color of the intermediate fluid to be monitored. The problem with this method is that the change in color must be *seen* before action can be taken; thus it is not as reliable.

J. Steam Jackets

When fluids that must be maintained at elevated temperatures are metered, steam jackets should be provided on the pump liquid end. The jacket shown in

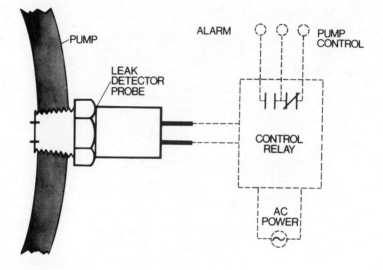

Figure 2.10. Leak detector system

Figures 2.11 and 2.12 consists of two sections, each with a carbon steel plate type heat exchanger in an aluminum alloy shell. The standard operating pressure is 150 psi at 600°F., and the unit may be field- or factory-mounted.

K. Rate Timers and Counters

For batching applications, metering rate timers and counters are used to totalize pump pulses and to shut off the flow when a specific limit has been reached.

L. pH Analyzers/Controllers

In closed-loop systems, pH analyzers are used to continuously monitor the process stream. Controllers adjust metering pump output in response to the analyzer signal. Additional information is provided in Chapter 5, Metering Pump Control Systems.

M. Feed System Accessories

Many metering pump systems include tanks to store chemicals that require continuous agitation. These tanks must be equipped with agitators. Dissolving

Figure 2.11. Two-section steam jacket prior to assembly.

Figure 2.12. Two-section steam jacket as assembled.

baskets may also be provided to permit addition of dry chemicals to feed system tanks. Floats are required when oxygen scavenging chemicals (e.g., hydrazine) are stored. Low-level alarms may also be included. A water treatment system incorporating some of these accessories is shown in Figure 2.13. Refer to Chapter 4, Chemical Feed Systems, for additional information.

II. CAPACITY CONTROLS AND CONTROLLERS

In addition to the preceding options, metering pumps may also incorporate electric and pneumatic capacity controls and controllers. Application of these devices in open, closed and open/closed loop process control systems is described in Chapter 5, Metering Pump Control Systems.

A. Electric Capacity Controls

To be suitable for use in automated processes, metering pumps require a method for automatic adjustment of stroke length (flow rate) in accordance with manual or electronic input signals. This control is achieved through a combination of two separate devices—an actuator and a controller (Figure 2.14). The actuator, which mounts on the pump body (Figure 2.15), contains a reversible motor to drive the pump capacity adjustment. The controller, which can be mounted remotely, contains electronic circuitry to accept an input signal and to command the actuator.

The types of controllers most frequently used with metering pumps include remote manual control with indication (RMI), proportioning relay with milliampere instrument follower control (PRRA-MA), and automatic milliampere instrument follower including remote manual control with indication (AMI) configurations. A brief description of each follows:

1. RMI—This is a remote manual control station with a control panel that indicates the output of the pump as a percent of stroke. The RMI controller permits manual adjustment of pump capacity from a remote location. Operation involves reversing switches, closed by a spring-return toggle switch, which drives the actuator to its various positions, Balanced bridge circuitry is not employed, but a feedback signal from the actuator adjusts a meter in the control station to indicate percent of full stroke length. The toggle switch is held in position until the meter indicates desired capacity.

2. PRRA-MA—This model contains a proportional bridge relay with ratio adjustment. It is used for automatic control of stroke length based on feedback from a process instrument that has DC milliamp output

(Figure 2.16). The milliamp rating of the controller must agree with the applied DC input signal.

3. AMI-MA—This unit combines the features found in RMI and PRRA-MA controllers; thus, it can be used for manual or automatic control of stroke length.

Additional information concerning application of electric capacity controls is provided in Chapter 5, Metering Pump Control Systems.

B. Pneumatic Capacity Controls

Pneumatic capacity controls are generally specified for hazardous (e.g., explosive) environments in which a pneumatic instrument signal is available. Controllers may include manual remote control, which typically provides a 3 to 15 psi air signal, or an electro-pneumatic relay for current-to-pressure (I to P) conversion. The former is adjusted by an operator; the latter permits automatic process control by converting a milliamp signal (e.g., from a flowmeter) to an air signal.

Most manufacturers offer pneumatic controls that automatically adjust the required stroke length (capacity) from 0 to 100% as a linear function of the applied instrument air signal. In many applications, the full-rated capacity of the pump is not required. It is desirable, however, that some means be available for trimming the maximum pump delivery at maximum signal conditions to the value needed for the particular application and for maintaining this delivery/signal ratio over the complete signal range. A ratio control adjustment in the instrument signal circuit is usually provided for this purpose.

III. MOTOR SPEED CONTROLS

Automatic variable-speed motor controls are generally used when applications require the range of metering pump delivery to correspond to the range of process signal variation and the ability to independently set or adjust maximum pump delivery.

New AC variable-speed drives entering the market are suitable for use in conjunction with metering pumps. These units contain micro-computer processing chips with memory that is pre-programmed to control and monitor all internal converter functions. Infinitely variable motor speed is claimed with constant torque over the full operating range.

Regardless of the type of variable-speed drive used, it is important to remember that a starting torque overload of 300% is typical in metering pumps, and that turn-down is limited by a minimum effective piston stroking rate of 15-20 strokes per minute.

Figure 2.13. Chemical feed system for water treatment applications including dissolving basket, agitator, and low level alarm.

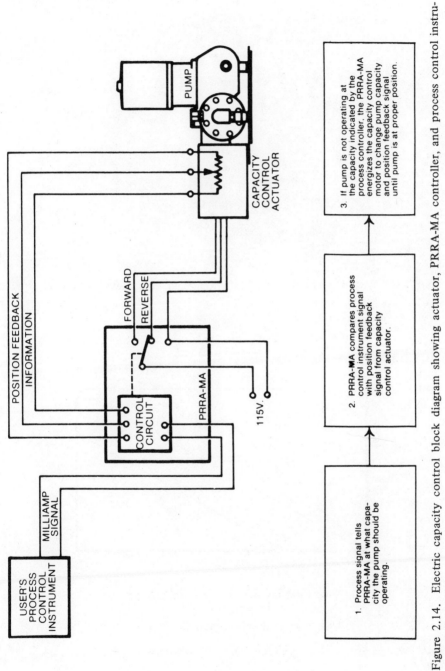

Figure 2.14. Electric capacity control block diagram showing actuator, PRRA-MA controller, and process control instrument interaction.

Figure 2.15.   Electric capacity control actuator mounted on metering pump.

PRRA-MA AND AMI-MA

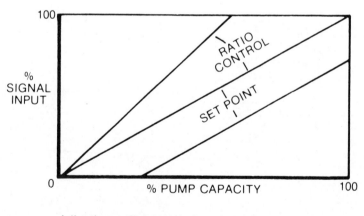

Adjusting ratio control changes slope of line about
the zero point.

Adjusting set point causes parallel displacement of line.

Figure 2.16.   Signal input versus pump capacity–PRRA-MA and AMI-MA controllers.

DC variable-speed drives employ silicon controlled rectifier (SCR) electronics and are compatible with resistance slidewire ma or DC voltage signals. Acceptable performance can be obtained over a 3:1 turn-down range, with extended range possible through use of a tachometer generator. Rated speed is difficult to maintain beyond 3:1 turn-down because of load changes arising from inherent metering pump sinusoidal torque characteristic. (A more complete explanation of turn-down ratios is provided in Chapter 5, Section 2A.)

Also, DC motor brushes constitute a hazard in explosive atmospheres. Other automatic control techniques should be considered under such circumstances.

Variable-speed controls are not generally used if zero capacity is required. Automatic capacity controls are sometimes used in conjunction with variable-speed motor controls to extend the linearity of pump delivery. This technique is discussed in Chapter 5, Metering Pump Control Systems.

3

SELECTING METERING PUMPS

In any application, the metering pump selected must provide the required combination of accuracy and reliability. Of course, there are other factors as well, ranging from automatic control options to proper sizing for the purpose of minimizing power consumption. Many of these considerations are covered in this and subsequent chapters.

Metering pumps resemble other engineered products in that many unique elements are reviewed when matching a pump to a specific application. Metering pumps, therefore, are not purchased by thumbing through a catalog—many factors must be considered, including:

1. Performance characteristics.
2. The parameters of the application.
3. The manufacturer's engineering capability.
4. Lead time for delivery.
5. Reliability of the design.
6. Price—first and later costs.
7. After-sales service.
8. Warranties.
9. Availability of spare parts.

These factors are the most important to any engineer who specifies a metering pump. Additionally, pumps should be sized as closely as possible to minimize equipment and operating costs.

In-depth selection information is provided on the following pages. In addition, Appendix A summarizes selection criteria for packed plunger, disc diaphragm and tubular diaphragm designs.

I. PERFORMANCE CHARACTERISTICS

Regardless of the type of metering pump employed, the metering function is accomplished by the pump's liquid end, where the reciprocating action of a plunger directly displaces the process fluid or imparts flexure in one or more diaphragms that displace the fluid. With the aid of one-way check valves, this action creates a pulse of accurately metered fluid.

As mentioned previously, a certain amount of packing leakage is necessary with the packed plunger design. In addition, friction causes the packing to wear, increasing leakage and producing a proportional decrease in volumetric efficiency. Thus, a certain amount of maintenance should be expected, including packing lubrication, adjustment and/or replacement.

Packed plunger pumps are not suited for handling fluids for which leakage cannot be tolerated. They are recommended, however, for high temperature fluids or applications in which high discharge pressures are required.

Most metering pumps in use today have diaphragm liquid ends. This type of pump provides an effective solution to process fluid leakage problems normally associated with the packed plunger liquid end class, because the diaphragm acts as a positive barrier between the plunger and the process fluid. See Chapter 1 for a discussion of various configurations of diaphragm liquid ends available.

II. APPLICATION PARAMETERS

A. Chemical Characteristics

Once the basic configurations are known, the next step is to define the parameters of the application and compare them with the performance characteristics of the available equipment. Parameters should include both chemical and physical aspects of the application as well as hydromechanical limitations of each type of metering pump.

Chemical features that must be considered include compatibility with corrosives, solvents, oils, liquified gases, radioactive fluids, hazardous or explosive fluids and food products or additives. Table 3.1 is a broad-brush guide to performance of the major liquid end types with various types of solutions.

1) Corrosives

All types of diaphragm pumps can be used to handle corrosives; however, the flow-through tubular diaphragm configurations are particularly well suited because they have minimal wetted surface area. Packed plunger pumps should not be employed due to the possibility of damage to the pump or surrounding area by leaking chemicals. For the same reason, packed plunger pumps are not a wise choice where toxic or explosive fluids are involved. These types of fluids are best suited for diaphragm pumps due to the positive membrane barrier.

TABLE 3.1

How Metering Pumps Stack Up to Common Pumpables

Type of Solution	Pump Type			
	Packed Plunger	Mechanical Diaphragm	Disc Diaphragm	Tubular Diaphragm
Acid	P	E	E	G
Slurry	E	E	F	E
Viscous	E	E	F	E
Polymer	E	E	F	E
Adhesives	E	E	F	E
Greases	E	E	F	E
Oils	E	E	G	E
Synthetics	E	E	E	G
Liquified Gases	P	P	E	P
High Temperature	E	G	G	F
Hazardous Liquids	P	E	E	E
Radioactive	P	G	E	F
Sterile	F	E	E	E

E – Excellent
G – Good
F – Fair
P – Poor

2) Solvents

Teflon disc diaphragms are well-suited for pumping solvents. Pressure drop through contour plate holes is not an important factor, because solvents have low viscosities. Conversely, many elastomeric tubular diaphragms are attacked by solvents. A compatible diaphragm material should be specified when employing this type; however, such a material can increase cost appreciably. Similar precautions should be taken when using mechanically actuated diaphragms, as many employ plastic check valves and other wetted components that are susceptible to solvent attack.

3) Oils

Compatibility with elastomers may again cause problems. However, where no compatibility problem exists, the flow-through tubular diaphragm is particularly well-suited because of the high viscosity of most oils.

4) Liquified Gases

Very low temperatures are associated with liquified gases; thus, remote, double diaphragm liquid ends are recommended. Remote diaphragms eliminate expansion and contraction of pump parts caused by large temperature differentials and facilitate maintenance of the temperature of the process fluid.

An unusual liquid end configuration has evolved for handling liquified natural gases. Known as the "cold pump" it features a remote, floating pump head. A duplex version is shown in Figure 3.1.

5) Radioactive Fluids

Radioactive fluids are best handled with highly specialized metallic diaphragm pumps, because Teflon and most elastomers are incompatible with such fluids.

6) Hazardous or Explosive Fluids

Virtually any kind of diaphragm pump is suitable, but the twin diaphragm pumps, with their intermediate fluid conductivity sensing capability, provided improved protection. Explosion-proof motors and controls are strongly recommended in these applications.

7) Good Products or Additives

Consideration should be given to the eventual need for sterilization techniques and resultant high temperatures. Due to their inherent high-temperature suitability, Teflon disc diaphragm liquid ends with stainless steel wetted parts are the best choice.

B. Physical Characteristics

Physical factors that must be considered include pump capacity, discharge pressure, viscosity, temperature, slurry settling and net positive suction head.

1) Capacity

Capacities above 750 gph are typically handled by packed plunger or tubular diaphragm. Disc disphragm pumps become impractical at high capacities because of disproportionate size and resultant cost of the liquid end. This design

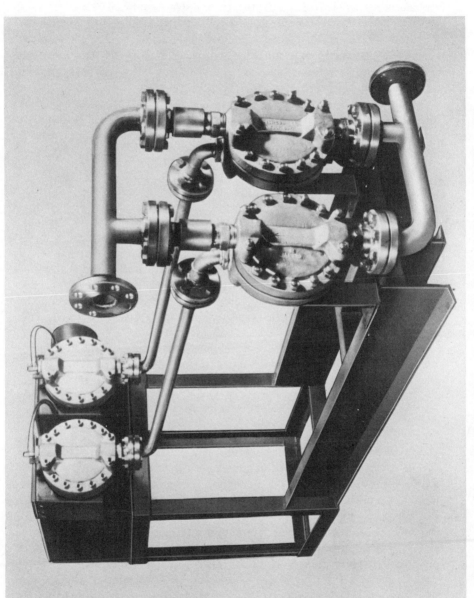

Figure 3.1. Remote, floating pump head eliminates problems caused by large temperature differentials.

characteristic arises from the limitation on diaphragm flex or travel and the requirement for very large diameters to achieve displacement volume.

Moderate flows—from 1 to 750 gph—can be handled by hydraulically balanced, disc diaphragm pumps. If other characteristics indicate a balanced diaphragm type liquid end, but required capacity is above 750 gph, consideration should be given to multiplexing diaphragm type liquid ends. Mechanically actuated diaphragm pumps can deliver up to 75 gph.

Low flows (below 1 gph) are best handled by packed plunger pumps since the "unswept" volume can be held to a minimum (high volumetric efficiency). Capacity drop-off at high pressure is thereby prevented and excellent repetitive accuracy maintained.

2) Discharge Pressure

High pressure metering can best be handled with packed plunger type pumps. The simple nature of the design (i.e., a piston reciprocating within a bore) facilitates pressure generation of up to 30,000 psig. Balanced diaphragm pumps incorporate a flexible membrane, and are limited to the pressures that such membranes can withstand when driven against supporting plates with flow-thru holes. Diaphragm pumps containing Teflon disc material can withstand pressures up to 4,000 psig, while tubular diaphragm pump types are typically limited to 3,000 psig or less.

Mechanically actuated diaphragm pumps have a large, unsupported membrane area exposed to a pressure differential and are generally limited to 150 psig. The preceding capacity and pressure limitations are summarized in Table 3.2.

3) Viscosity

Successful viscous fluid pumping depends only in part on pump characteristics. The more important consideration in high viscosity applications is the nature of the suction piping system. From a practical standpoint, "adequate flooded suction" as opposed to simply "flooded suction" is essential to proper metering, and is an indication that the suction plumbing will deliver fluid at a rate equal to the peak flow rate of the pump. Since most metering pumps draw in liquid in a periodic fashion, this means that free flow at the pump suction port must be approximately π times rated maximum flow (see Chapter 6, Installation).

Proper free-flow delivery can be achieved by elevating or pressurizing the supply vessel or through use of a small gear pump to feed the metering pump. Other parameters that should be considered include minimizing suction system pressure losses through short piping runs and use of large diameter pipe.

Pump selection should be based upon minimizing pressure losses during the suction stroke. Single-ball check valves are preferred to double-ball valves,

TABLE 3.2

Typical Metering Pump Specifications*

Pump Type	Maximum Pressure (psig)	Maximum Capacity (gph)
Mechanically Actuated Disc Diaphragm	150	60
Hydraulically Actuated Disc Diaphragm	4000	750
Tubular Diaphragm	3000	300
Tubular/ Disc Diaphragm	1000	750
Double Tubular Diaphragm	300	250
Double Disc Diaphragm	4000	600
Packed Plunger	7500	1500

*Individual manufacturers may offer specialized designs that exceed pressure or capacity values shown. Due to limited source selection, however, performance extremes are not listed.

and liquid ends without restrictive support plates (including packed plunger and tubular diaphragm types) meet this criterion. Other important characteristics include operation at low stroking speeds and selection of large-diameter plungers to reduce viscous drag losses.

Assuming proper system design, including pump type selection, liquids with viscosities up to 30,000 cp can readily be handled by reciprocating metering pumps.

4) Temperature

The ability of a given pump design to withstand high temperature is a function of seal, packing and diaphragm material performance as well as lubricant stability. Once again, a packed plunger design is best suited since diaphragms are not used, high-temperature seals and packing are readily available and gear lubricants

TABLE 3.3

Practical Temperature Limits for Diaphragm Materials

Material	Lower Limit − °F	Upper Limit − °F
FLU (fluorinated hydrocarbon)−Viton	−20	300
TFE (tetrafluoroethylene)− Teflon	−60	250
EPDM (ethylene propylene terpolymer)	−20	175
CSM (chlorosulfonated polyethylene)−Hypalon	−20	175

Temperature limits for various diaphragm materials depend on a number of factors including support plate hole diameter, desired flexure life and stiffness.

are removed from the immediate vicinity of process-wetted parts. Also, stuffing box cavities can be filled with a suitable fluid and cooled by tubing coils connected to a chilled water source.

Typical temperature limits (process fluid) are 400°F for packed plunger, 250°F for TFE disc diaphragm and 175°F for tubular diaphragm types. Table 3.3 provides practical temperature limits for the various diaphragm materials. Individual manufacturers should be consulted, however, to determine if synthetic lubricants, special materials of construction, ambient cooling and/or remote check valves or diaphragm heads can be used to extend temperature limits for a given design.

Low-temperature fluids such as liquified gases are best handled by pumps with remote disc diaphragms. Teflon is generally suitable down to -60°F and the distance of the liquid end from the housing reduces temperature effects on gear box lubricants.

Another point to consider—the friction associated with the packed plunger liquid end may result in heat that can vaporize the process liquid, particularly as pump chamber pressure approaches the vapor pressure. Liquids handled near their boiling points should be metered using remote diaphragm heads.

5) Slurries

The tubular diaphragm and packed plunger pump configurations are best suited for handling slurries due to minimal flow restriction. However, extremely fine dispersions and low concentrations by weight with solids up to 200 mesh in size can also be successfully handled by disc diaphragm pumps. More concentrated slurries are prone to settle between the contour plate and the diaphragm in typical disc diaphragm pumps.

When packed plunger pumps are used for slurries, an internal flush should be specified to keep particulate away from the packing-plunger contact interface. A through flush should be installed when contamination of the process liquid with the flushing agent is not desirable. Also, ceramic or hardened steel plungers should be used to prevent scoring due to abrasion.

Check valves with hardened balls and seats should always be specified when abrasive slurries are pumped. In addition, if solids are apt to come out of suspension or settling is likely to occur during system shutdown, an appropriate flushing provision (i.e., three-way valves) should be included upstream and downstream of the pump. Similar consideration should also be given to the rest of the system piping.

6) Net Positive Suction Head (NPSH)

By calculating NPSH, the manufacturer verifies that the pump will have enough suction head pressure to deliver its proper rated flow. Insufficient pressure may result in pump cavitation or starvation in the liquid end, destroying the metering action of the pump.

If initial user calculations indicate unacceptable suction head conditions, the following steps may be taken to reduce suction head losses:

1. Add an accumulator to dampen line flow pulsations and reduce suction pipe flow and acceleration losses.
2. Increase suction line diameter.
3. Use the next larger check valve size.
4. Shorten the pipeline to reduce flow losses.
5. Pressurize the supply tank, raise the supply tank or lower the pump to increase suction line pressure.
6. Heat the process liquid to lower its viscosity.
7. Chill the process liquid to reduce its vapor pressure.

A method for calculating NPSH is discussed in Chapter 6, Installation, and detailed in Appendix C.

III. ENGINEERING CAPABILITY

The preceding types of liquid ends are available off-the-shelf from most manufacturers, but non-standard designs are often dictated by unusual process conditions.

```
Cartridge Size   1.250              File#._____         test___
Seat Diameter    0.658              Pump Model. Milroyal A/B
Ball Diameter    0.875              Liquid End.   Packed Plunger
                                    Pipe ID  1.049
                    'NPSH' DATASHEET


1. System Data Required:
    a. Liquid Viscosity.....................   20(centipoises)
    b. Liquid Specific Gravity..............    1.00(no units)
    c. Liquid Vapor Pressure................    0.10(psia)
    d. Pump Flow Rate....................... 120.00(gph)
    e. Pump Stroke Speed....................  142(spm)
    f. Plunger Diameter.....................    1.75(inches)
    g. Nominal Suction Pipe Diameter........    1.00(inches)
     * (L/D Ratio)......................... 100.00
    h. Equivalent Fitting Length............    9(ft)
    i. Actual Suction Pipe Length...........    4(ft)
    j. Equivalent Suction Pipe Length.......   13(ft)
    k. Vertical Distance of Lowest Liquid
       Level in Supply Resevoir
       Above Pump.........................+    2.0(ft)
    l. Pressure on Liquid Surface in
       Supply Resevoir.....................   14.7(psia)


2. To Determine Viscous Drag Pressure
   Losses:
    c. Total Suction Pipe Flow Loss.........    0.4(psi)
    d. Suction Ball Check Cartridge Loss....    0.4(psi)
    f. Operating Saftey Margin..............    3.0(psi)
    g. Total Viscous Drag Flow Loss.........    3.8(psi)
    i. Elevation Head Pressure..............    0.9(psi)
    j. Total of Pressure Sources............   15.6(psia)
    k. Minimum Pressure on Liquid
       in Liquid End.......................   11.8(psia)


3. To Determine Liquid Acceleration Loss:
    d. Acceleration Loss....................    2.5(psi)
    e. Operating Saftey Margin..............    3.0 (psi)
    f. Total Acceleration Loss..............    5.5(psi)
    h. Minimum Pressure on Liquid
       in Liquid End.......................   10.1(psia)
```

Figure 3.2. NPSH Printout provides computerized confirmation that a specific pump will perform adequately in a specific application.

Where such a design is required, a manufacturer with extensive engineering capability is best equipped to recommend the optimum design based on parameters supplied by the user.

Exotic liquid end metallurgies or hardened titanium/tungsten carbide check valve seats, ceramic balls and plungers and diaphragms molded from alternate materials typify non-standard components that may be recommended. (Standard disc and tubular diaphragms are generally made of Teflon and elastomers, respectively.) Other non-standard accessories include pulsation dampeners, stroke pacers and counters, flowmeters, shut-off valves, remote diaphragm assemblies, annunciators, relays or other special controls (see Chapter 2, Metering Pump Accessories).

Due to the sinusoidal flow characteristic and other dynamic operating variables that must be considered in selecting a metering pump, specification is often best determined in concert with the manufacturer. The NPSH printout shown in Figure 3.2 is an example of performance confirmation data available from a manufacturer during the pump selection process.

IV. LEAD TIME FOR DELIVERY

Delivery is an important consideration when purchasing any piece of equipment. Metering pumps are no exception—sufficient lead time for delivery must be built into the overall schedule.

Many smaller manufacturers subcontract component machining, an added step that often lengthens delivery time. Assurance of on-time delivery is improved by selecting a company with total in-house manufacturing capability.

Standard pumps should be ordered two to four weeks ahead of required delivery date; non-standard designs may require additional lead time. If a non-standard design or one with exotic metallurgies is involved, the order should be placed at least four to six weeks ahead of required delivery date. Due to lower cost and ready availability, a standard pump should be specified where possible.

V. RELIABILITY

A paramount consideration is the reliability of the pump selected. Because there is no specific system for judging reliability in the metering pump industy, the user is forced to consider less-tangible factors, such as:

1. The length of the warranty.
2. The reputation of the manufacturer.
3. The reputation of the product.
4. Prior experience with similar products.
5. Soundness of the design.

Metering pumps are often used under extremely harsh service conditions where proper operation is vitally important; for example, a pump used for continuous addition of catalysts in a plastic manufacturing process. Trouble-free operation is imperative to maintain both production and product quality. In many waste treatment applications, proper metering of additives prevents violation of federal standards. Similarly, unacceptable stack emissions and possible

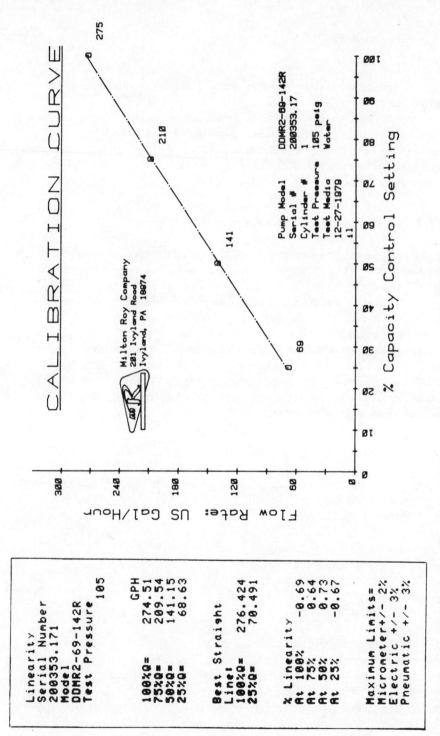

Figure 3.3. Automated printout of pump performance under dynamic operating conditions.

52

EPA fines are avoided through use of metering pumps for addition of combustion enhancing chemicals. Other critical applications include maintenance of water quality in boilers to help prevent costly tube fouling and addition of biocides to cooling tower water.

Due to the high stakes involved in many applications, the importance of a time-proven design cannot be overestimated. Use of cast pump housings and TEFC motors are recommended, particularly for outdoor installations and long-term service where dust, heat, and moisture may be encountered.

The effects of outdoor installation present unique problems to exposed plastic components, including ultraviolet degradation, leeching out of plasticizers, and the deleterious effects of ozone. If other considerations make plastic pumps necessary for outdoor use, impervious materials such as polyethylene with UV stabilizers should be specified, and the pump should be shielded from direct sunlight.

To assure reliability, the user should select a metering pump based on maximum operating pressure and capacity. He should also verify that the pump has been hydrostatically tested to 1-1/2 times the rated pressure.

Users in the chemical processing industry often specify a pump that operates at 70% of its maximum capacity. Due to the fluid dynamics of many systems, this provision is necessary to allow for last minute changes. Because electric motors have been singled out as major energy consumers, many experts predict the practice of oversizing pumps will not continue. Users in the CPI will be forced to pay closer attention to exactly sizing a pump for its application. (See Chapter 9, Energy Considerations, for additional info.)

When pumps are exactly sized, the capacity must be carefully derated as pressure increases. Output drops off slightly as a function of the compressibility of the fluid, which increases with pressure. Capacity is typically derated 1% for every 100 psi increase in discharge pressure beyond 1,000 psi. The capacity of a pump operating at 3,000 psi should therefore be derated 20%, because it will only provide 80% of its rated flow under actual conditions. A calibration curve as exemplified in Figure 3.3 should be specified to confirm performance under simulated operating conditions.

VI. FIRST AND LATER COSTS

Domestically manufactured metering pumps are all competitively priced. Thus, the most reliable, robust design should be sought. Imports, which are finding their way into the American market, are generally more costly; service, delivery and parts availability can also present problems in this case.

In general, standard models are less costly than non-standard models, and should be utilized where possible. A very wide range of metering pumps is available right off the shelf, with a myriad of materials, sizes and pressure ranges.

Due to the nature of most metering pump applications, first cost is generally not a prime consideration. It is more important that the design meets other technical requirements. There are other reasons the user should remain cost-conscious, however.

For example, although packed plunger pumps are generally offered at lower first cost than pumps with diaphragm liquid ends, routine maintenance is required in the form of packing lubrication, adjustment and replacement. Although a greater initial outlay is required for diaphragm pumps, payback from maintenance savings can be significant.

Moreover, a metering pump may require not only mechanical maintenance, but also electrical or instrument maintenance. An unreliable design that breaks down may require a whole range of costly services. Therefore, a robust, reliable pump, though initially more expensive, can bring about sizable savings in the long run.

VII. AFTER-SALES SERVICE

The user should purchase from a company with a factory service organization to permit consultation when after-sales problems arise. These could include field performance difficulties or the need for retrofit, such as addition of an automatic capacity control. The service organization should also be available to assist with installation, start-up, testing, and calibration.

VIII. WARRANTIES

A pump warranty provides another method for determining the reliability of a given pump design. The period of time it covers not only ensures the user against additional cost due to breakdown, but also provides up-front evidence of the reliability of the design.

A 90-day or six-month warranty indicates the pump may perform well, but with very little safety margin. If the rated pressure is exceeded even slightly, the pump will most likely fail in short order. Conversely, a one- or two-year warranty indicates the likelihood of a rugged design that will take much more abuse.

Prior to purchasing the pump, the engineer should seek a model with a one- or two-year warranty, particularly on the pump drive train.

The warranty should also extend to all service calls that are not related to defects caused by misuse, abuse, or improper application or operation. Damage caused by wear or chemical action is generally excluded.

IX. SPARE PARTS

Because many unique design elements are employed in metering pumps, availability and delivery of spare parts is an important consideration. The manufacturer should provide a list of recommended spare parts—those most likely to be needed due to rapid wear, chemical action or other factors.

Key items include such high-wear items as check valve balls and seats, packing, lubricants and any special tools that might be required. Immediately after the pump is delivered, the user should order several sets of spare parts. The cost per set is typically very small, ranging from 5 to 10% of total purchase price.

X. AVAILABILITY OF PACKAGE SYSTEMS

An important consideration in reducing system design and construction lead time is use of self-contained, packaged chemical feed systems. Various manufacturers offer skid-mounted packages containing all necessary equipment, including pumps, tanks, piping, hydraulic accessories and electrical controls. Standard systems are readily available with a variety of accessories that allow the customer to easily tailor equipment to meet specific requirements. Complex systems can also be designed and fabricated to meet exact specifications. The following chapter provides detailed information on custom-engineered chemical feed systems.

4

CHEMICAL FEED SYSTEMS

Chemical feed systems generally require a pump whose output can be varied through manual or automatic adjustment of the displacement mechanism. Due to the wide nature of the fluids that are handled, which range from abrasive slurries to corrosives and the high discharge pressures that are commonplace, reciprocating metering pumps are almost always used.

Chemical feed systems are self-contained units that store, mix and dispense chemical solutions. As generally defined, a chemical feed system consists of at least one storage or mixing tank, one or more metering pumps and related equipment. The entire system is usually mounted on a skid or packaged as an integral unit to allow direct installation into a larger system or process.

Although standard chemical feed systems are available for certain water and wastewater treatment applications, more complex tasks—treatment of boiler water and feedwater and injection of various petroleum additives, for example—require custom-engineered systems. As a result, thousands of one-of-a-kind systems are presently in operation throughout industry (see Figures 4.1, 4.2, 4.3, and 4.4). A few of the chemicals handled by these systems are listed in Table 4.1.

Because most chemical feed systems are unique, this chapter is primarily concerned with design methodology. In order to approach a complex subject in an orderly fashion, it is necessary to begin with preliminary layout and proceed through skid design, tank selection and sizing, and piping, control panel and hydraulic accessory selection. The guidelines presented in Chapter 3, Metering Pump Selection, also apply to selecting metering pumps for chemical feed systems.

I. PRELIMINARY LAYOUT

Unless the system is a standard design (such as that shown in Figure 2.11) the first step is to assemble a preliminary layout to provide a general idea of the overall size and location of the equipment. From the preliminary layout, a second drawing showing electrical and mechanical connection points and size and location of all equipment should be prepared. Next, a piping drawing is needed that includes size, connection technique and location of all pipe and

Figure 4.1. Custom-designed chemical feed system with control panel.

Figure 4.2. Chemical feed system showing calibration gauges that accurately verify pump output. Gauges are installed in the suction line and vented to permit drawdown.

Figure 4.3. Chemical feed system complete with master control panel is shown. Various analyzers, annunciators, recorders, controllers and switches are visible— circuit breakers, relays, motor starters and other control elements are housed inside the panel.

fittings. From these drawings, a fairly accurate bill of materials (BOM) can be assembled and ordered. Prior to assembling complex systems, detailed final fabrication drawings that include tank and skid details may be necessary. Figures 4.4-4.9 illustrate the progression from preliminary layout to final fabrication drawings.

Once the system has been built on paper, actual construction begins. Before even the most basic layout can be completed, however, a thorough understanding of the available options and function of each component is necessary.

II. SKID DESIGN

Skids are recommended if the system will be moved after construction, or as a means of attaching system components. The skid is a skeletal frame on which all components are mounted. Skids should be constructed of structural steel, and designed for maximum strength to avoid induced loads on piping and joints. Skids may be constructed of hot or cold rolled steel in combination with structural members such as channel and angle iron.

Skids are sometimes plated with various materials to resist corrosion and can include sumps and other provisions to carry off drainage to safe locations.

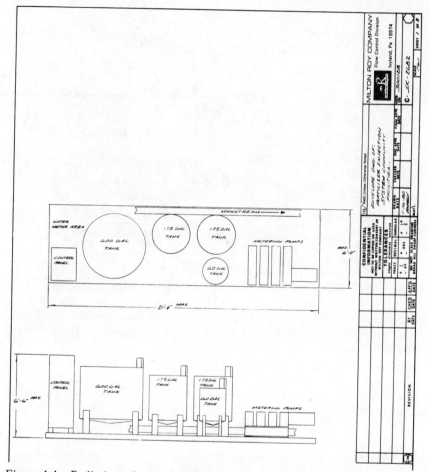

Figure 4.4. Preliminary layout for fertilizer injection chemical feed system.

These types have deck surfaces that are sloped toward the drains to assist in carry-off of spilled or washdown liquids.

Ladders with railings are often necessary to permit manual addition of chemicals to tanks by employees. Ladders may be permanently attached or removable. Large systems often incorporate walkways with safety railings to facilitate access to tanks.

Skids are normally welded, one-piece assemblies. Larger systems may require welded modular sections that are bolted together. Chemical-resistant coatings should be applied where necessary, and lifting lugs should be provided if the skid will be moved. Provisions for bolting the skid to a floor or concrete pad may be necessary in some systems.

TABLE 4.1

Chemicals Frequently Handled in Packaged Systems

Chemical	Function	Application
Alum	Coagulant	Potable/Waste water
Amines (various)	Neutralize acid; corrosion inhibitor; coagulant	Boiler water
Ammonium hydroxide (Aqueous ammonia)	Neutralize acid	Boiler water
Calcium carbonate* (Lime slurry)	Neutralize acid	Waste water
Calcium hyroxide* (Slaked lime)	Neutralize acid	Waste water
Chelants (various)	Remove metal ions	Boiler/Waste water
Ferric chloride	Coagulant	Waste water
Fuel oil additives (various)	Improve combustion efficiency	Fuel oil
Hydrazine	Oxygen scavenging; Corrosion inhibitor	Boiler water
Lime slurries*	Neutralize acid	Waste water
Morpholine	Corrosion inhibitor	Boiler water
Sodium hydroxide (caustic soda)	Neutralize acids Provide hydroxyl ions	Boiler/Waste water Demineralizers
Sodium hypochlorite	Bateriocide	Potable/Waste water
Sodium phosphates*	Detergent; buffer; softener	Boiler water
Sodium sulfite	Oxygen scavenging	Boiler water
Sulfuric acid	Neutralize bases Provide hydrogen ions	Waste water Demineralizers

*Continuous packing flush recommended in packed plunger pumps.

ITEM	QTY	DESCRIPTION	MAT'L.
A	4	½" × ¼" 3000# THD BUSHING	316 SS
B	2	½" 3000# THD. UNION SEAT SAME MATL AS BODY	
C	4	½" 3000# 90° THD. ELL	
D	3	½" 3000# THD. TEE	
E	4	½" 600# THD BALL VALVE GRINNELL 1560 OR EQUIV.	
F	2	½" 150# THD BALL VALVE GRINNELL 1560 OR EQUIV.	
G	3	½" 600# ANSI. THD. RF. FLANGE	
H	2	¼" 3000# THD. TEE	
J	1	½" 3000# THD CROSS	↑
L	6	¼" SCH 80 THD. NIPPLE 2" LG.	304 SS
M	2	⅜" SCH. 40 THD. NIPPLE 2" LG.	
N	12	½" SCH 80 THD. NIPPLE 2" LG.	
P	2	½" SCH 80 THD NIPPLE 5" LG.	
R	2	½" SCH 40 THD NIPPLE 2" LG	
S	1	½ SCH 80 THD NIPPLE 10⅛" LG	
T	2	1" 150# THD. UNION SEAT SAME MAT L AS BODY.	
U	4	1" 150# 90° THD ELL	
V	3	1" 150# THD TEE.	↑
W	2	1" 150# THD 'Y' TYPE STRAINER SCHNITZER	316 SS
X	4	1" 150# THD. BALL VALVE GRINNELL 1560 OR EQUIV	316 SS
Y	1	1" 150# ANSI THD. RF FLANGE	304 SS
Z	2	1" × ⅜" 150# THD BUSHING	
AB	13	1" SCH. 40 THD. NIPPLE 2" LG	
AD	2	1" SCH 40 THD NIPPLE 4⅜" LG	↑
		ORDER ALL ITEMS MARKED THUS.	
REF	1	½" SCH 80 PIPE 1'-3½" LG	316 SS
REF	1	1" SCH 40 PIPE 1'-2¾" LG.	304 SS
		ALL CONNS ARE NPT.	

MILTON ROY COMPANY
Flow Control Division
Ivyland, Pa. 18974

USED ON 300212

C. 300212 -1 -001　　Ⓑ

Figure 4.5.　Typical bill of materials for chemical feed system.

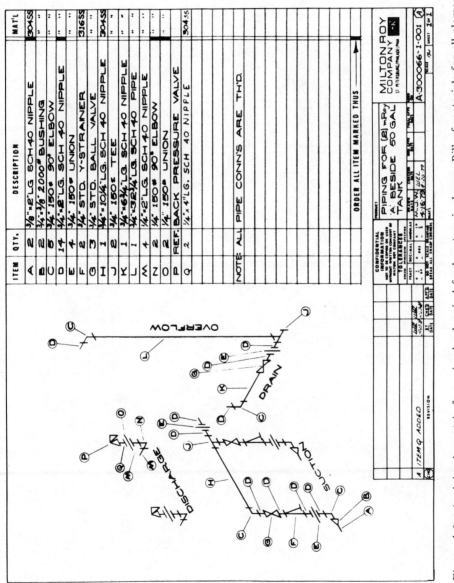

Figure 4.6. A piping isometric for a simple chemical feed system is shown. Bill of materials for all elements of the piping network is included.

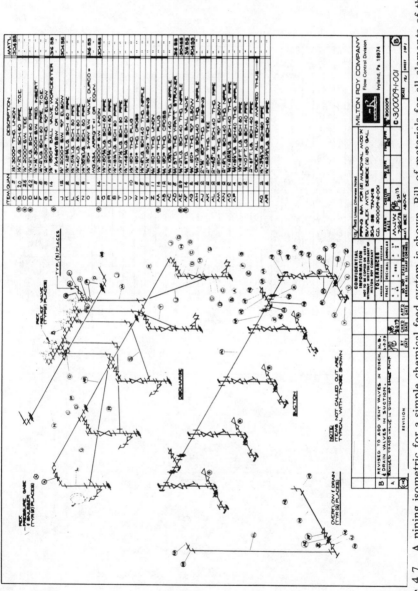

Figure 4.7. A piping isometric for a simple chemical feed system is shown. Bill of materials for all elements of the piping network is included.

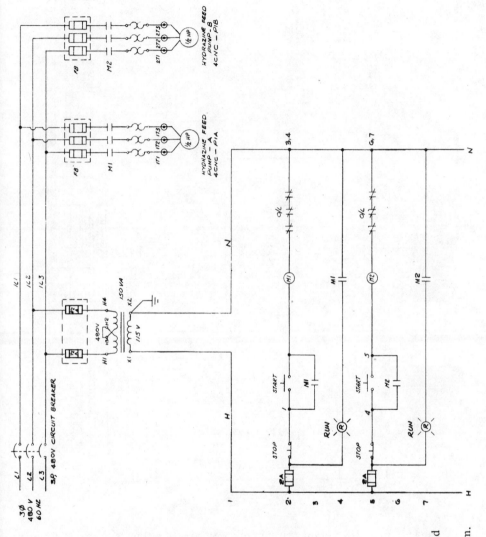

Figure 4.8. Electrical and mechanical drawing for hydrazine injection system.

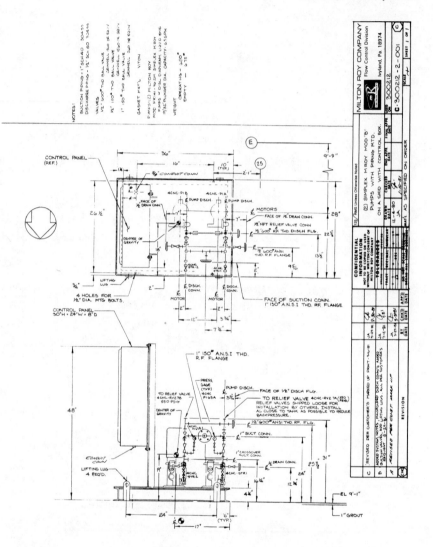

Figure 4.9.   Final fabrication drawing for chemical feed system with two metering pumps.

III. TANK SELECTION

The process liquid is stored and mixed in one or more tanks, each of which provides a continuous supply of its contents to the metering pumps. Each tank should be sized according to the required flow rate of its associated pump(s), thereby ensuring reasonable filling intervals.

Although tanks are generally mounted in a vertical attitude, horizontal tanks supported on saddles are sometimes used. Tanks should be supported by legs, which are in turn mounted to the skid. This helps improve suction conditions to the pump and facilitates washdown. Vertical steel tanks can be elevated on angle legs welded to the sides of the tank, while plastic tanks are inserted into a ring base arrangement that provides the necessary support.

Unpressurized storage tanks can be as large as 1,500 gallons, but the 50- to 500-gallon range is much more common. Compatibility with the chemical being stored is very important in selecting tank material, which can range from hot or cold rolled steel to stainless alloys. Compatibility can be enhanced through use of chemical-resistant coatings applied to the wetted tank surfaces.

Fiber glass tanks are recommended for extremely corrosive materials. All-plastic tanks may do just as well in certain cases, and should be considered from a cost standpoint. Plastic-lined tanks are generally used in higher pressure applications where all-plastic construction lacks sufficient strength and economic considerations override use of all-metal material. Normally, tanks should be covered with bolted and flanged covers that are hinged to facilitate access. Lightweight, removable covers may also be used.

Pressurized tanks are commonly used for ammonia service, with the ammonia concentrate added through a sparger. The pressure limit is, in general terms, inversely proportional to the tank's capacity. A 500-gallon tank, for example, is generally limited to about 150 psi. Pressurized tanks should be supported with saddles, and should include welded-on dome or flanged dish-shaped tops and bottoms to minimize stress levels. A flanged port should be included for access to the tank interior.

IV. TANK ACCESSORIES

Tanks that will hold chemicals requiring constant or intermittent agitation must have a provision for mounting an agitator and a penetration in the cover for the agitator shaft. Agitators may be fixed or bracketed for removal. Agitators consist of a shaft, propeller and electric or pneumatic motor. A bubbler-type agitator is sometimes used where the process fluid is compatible with air.

Propeller and shaft materials selected must be compatible with the stored chemical. To avoid the abrasive effects of certain chemicals, a protective coating should be applied to the propeller.

The rate of agitation and type of propeller depend on the nature of the fluid in the tank. For extremely viscous fluids, a very slow speed, right-angle agitator works best. Air-scavenging fluids require sealed tanks and agitator assemblies and slow agitation speeds to minimize atmospheric contact. With some air-scavenging fluids, such as hydrazine, a metallic or styrofoam float can be inserted in the tank to further reduce the surface area of the fluid.

Figure 4.10. Small chemical feed systems utilizing a small concentrate or shot tank for precise chemical batching.

Shot tanks are often used to add a concentrated solution to the main batch tank. The auxiliary shot tank is flange-mounted to the upper edge of the main tank (Figure 4.10). An agitator may be installed in the main tank to mix the concentrate as it is gravity-fed from the shot tank.

If dry chemicals in heavy granular or nugget form must be added to the main tank, a perforated dissolving basket may be required. The basket should be hung inside the tank, where circulating flow dissolves the chemical and carries it into solution. An agitator should also be included with this type of assembly.

Other tank accessories include calibrated gage glasses and various types of level controls. Gage glasses allow visual monitoring of the level of the fluid in the tank and are an inexpensive means of calibrating the system when the falling level is timed with a stopwatch. Three types of gage glasses are commonly employed: clear plastic tubing, exposed glass protected by guard rods, and armored assemblies. Armored gage glasses are recommended for hazardous or corrosive fluids, and contain shut-off ball valves that prevent loss of tank contents in case of breakage.

Tanks often include more sophisticated level controls that automatically activate relays and warning devices. Depending on the nature of the fluid and the desired expenditure, various types of electromechanical or capacitance level indicators can be used to provide continuous or point monitoring. One common combination involves high, low and low-low indicators—when the level reaches the high point, the pump starts. At the low point, the pump stops and at the low-low threshold an alarm is activated.

V. PIPING

A thorough knowledge of piping materials and techniques is also required. System piping must not only be matched with the chemical environment, but also the pressure, temperature and type of service. In addition to the information provided herein, additional information pertaining to sizing metering pump suction and discharge piping is provided in Chapter 6, Installation. Pipe material should be selected that is resistant to process liquid corrosion and galvanic corrosion at pump connection points. Inside diameters should match perfectly between connections, and all burrs, sharp edges and weld splatter should be removed from the lines. It is advisable to blow out or flush all lines before connection.

If no metal alloy is resistant to the process chemical, short runs of plastic piping may be used. If longer runs are required, or pressure is too high for plastic pipe, plastic-lined metal piping should be employed. Certain small-diameter lines (1 in. or less) are often fabricated from tubing, which is available in most metallurgies. When tubing is used, care should be taken to employ maximum bend radii.

TABLE 4.2

Pipe Pressure — Temperature Rating

The following format for calculating pressure — temperature rating for pipe is commonly used in Power Plant Piping Systems within the scope of Section I, Paragraph 104, of the Code for Pressure Piping, ANSI B31.1.0, and Section I on Power Boilers of the ASME Boiler and Pressure Vessel Code.

$$P = \frac{2 \, SE \, (t_m - C)}{D_o - 2y \, (t_m - C)}$$

t_m = Minimum pipe wall thickness in inches ($87\frac{1}{2}\%$ of nominal wall).
P = Maximum internal service pressure in PSI gage.
D_o = Outside diameter of pipe in inches.
E = An efficiency factor for longitudinal welded pipe. Use $E = .60$ for butt-welded ASTM A-120, Grade A. For seamless pipe, use a factor of $E = 1.0$.
S = Allowable stress in material due to internal pressure, at operating temperature, in PSI.

     (1) For ASTM A-120, A-53 and A-106, Grade A at –20 to 650°F, use 12,000 psi.
     (2) For ASTM 312, Grade TP 316 at –20 to 100°F, use 18,700 psi.
     (3) For ASTM 312, Grade TP 304 at –20 to 100°F, use 18,700 psi.
     (4) For materials other than those listed, use Table UHA-23 of Section VIII of the ASME Boiler and Pressure Vessel Code.

*C = Allowance for threading and structural stability.
    = .05 inches for under $\frac{1}{2}''$ pipe size.
    = .065 inches for $\frac{1}{2}''$ to $3\frac{1}{2}''$ pipe size.
    = .000 inches for $4''$ pipe size and larger.

*Note: No allowance has been made herein for corrosion or erosion. If such allowances are required as determined by the designer, they should be added to "C" in the formula and the reduced allowable pressure calculated thereunder.

y = A coefficient having values as follows:

Temp. °F	900°F and below	950°F	1000°F	1050°F
Ferric steels	.4	.5	.7	.7

Example

For $\frac{3}{4}''$ pipe, Sch 40, ASTM A-120, butt-welded and threaded, 200°F service temperature:

P =
t_m = $(.140 \times .875) = .1225$
D_o = 1.66
E = .6
S = 12,000 psi
C = .065
y = .4

$$= \frac{2 \, (12,000) \, (.6) \, (.1225 - .065)}{1.66 - 2(.4) \, (.1225 - .065)}$$

$$= \frac{828}{1.6} = 517 \text{ psi}$$

TABLE 4.2 *(continued)*

The following table covers a wide range of common pipe sizes, types and service temperatures as listed. For stainless steel seamless pipe, ASTM 312, TP 316 and ASTM 312, TP 304 at a service temperature of –20 to 100°F, use the table values times a multiplier of 1.56.

*Working Pressure for Steel Pipe — (–20 to 650°F)

Butt-welded threaded steel pipe ASTM A-120 or equal.			(1) Seamless threaded steel pipe ASTM A-106 & A-53 Grade A.		
Pipe Size	Sched. 40	Sched. 80	Sched. 160	Sched. 40	Sched. 80	Sched. 160
1/8	344	1260	N/A	574	2100	N/A
1/4	750	1568	N/A	1250	2614	N/A
3/8	661	1384	N/A	1102	2306	N/A
1/2	536	1169	1895	882	1947	3158
3/4	479	1014	1928	798	1690	3214
1	577	1067	1861	961	1779	3101
1-1/4	517	931	1443	863	1551	2405
1-1/2	482	874	1485	804	1457	2475
2	434	798	1554	724	1330	2591
2-1/2	554	932	1421	924	1554	2369
3	525	853	1411	875	1422	2351

Note: For ASTM 312 TP 304 & TP 316 at –20 to 100°F, use a multiplier of 1.56.
N/A — Not available.
*Abstracted from Section 1 of Code for Pressure Piping (ASA B31.1)
A safety factor of (4) is utilized.
(1) ASTM A106 Grade A up to 1-1/2", over 1-1/2", ASTM A53 Grade A used.

Pipe and tubing are commonly available in three schedules: 40, 80 and 160. The schedule selected depends on the pressure and temperature of the process fluid. Heavier-than-necessary schedules are often used to extend corrosion resistance, while lower-than-necessary schedules may suffice for pump suction lines. As a rule of thumb, the pressure limit is proportional to the schedule of the piping. Table 4.2 outlines pipe pressure rating systems recommended by the American National Standards Institute (ANSI) and the American Society for Testing and Materials (ASTM).

There are four major types of pipe connections—threaded, butt welded, socket welded and flanged. No matter which type is chosen, threaded or flanged connections should be provided at connection points to tanks, pumps and other accessories for ease of removal.

A. THREADED CONNECTIONS

Fatigue failure due to cyclic vibrations, pulsations, mechanical stress and galling during assembly and disassembly is most likely when threaded connections are employed. They are more apt to leak and have poorer flow characteristics than other types.

B. SOCKET-WELDED CONNECTIONS

These are often required by construction codes in refineries and throughout the oil industry. Assembled properly, such connections are leak-tight and are advantageous in vibrating systems. Socket-welded connections also provide smoother flow characteristics than threaded types. Because an extra coupling is required at each joint, however, material and labor costs increase. Disassembly is also difficult. During assembly, a 1/16-in. gap should be left in each joint to compensate for expansion.

C. BUTT-WELDED CONNECTIONS

Because they offer the least resistance to flow, butt-welded connections are normally used in pipes with larger diameters. While butt welding provides a reliable, leak-proof joint, the assembly process is very labor-intensive due to the difficulty of alignment. Disassembly is also difficult.

D. FLANGED CONNECTIONS

Flanged connections require a gasket that fits between two bolted flanges. Although not as economical, flanged connections are more reliable than threaded types. They should be used wherever assembly or disassembly is required, particularly in hazardous locations or where the process fluid might block the lines. In all cases, the gasket material must be compatible with the process fluid.

VI. CONTROL PANELS

The control panel is the brain of the entire system—emanating from it are power leads for motors and control wiring for all electronically operated components. Complexity can range from simple junction boxes to master panels containing analyzers, controllers and recorders (Figures 4.3 and 4.4). The actual design of control panels is beyond the scope of this work and is best left to an instrumentation engineer.

Chemical feed system control can range from manual to entirely automatic. Where it is acceptable to run conduits to all electrical equipment in very simple systems, the control panel might not be necessary at all. From pocket-sized to room-sized, control panels may contain anything from simple motor starters to mini-computers.

Control panel housings can be fabricated to explosion-proof, weather-proof or other NEMA classifications. Small control panels can be mounted directly to the side of the tank; larger units should be mounted on brackets. When size makes it impractical to mount the panel directly on the tank, it can be foot-mounted on the skid. The control panel should be readily accessible, and may be remotely located in hazardous areas.

VII. HYDRAULIC ACCESSORIES

It is often necessary to include the following hydraulic accessories with chemical feed systems:

1. Calibration gages to check metering pump accuracy. Although gauge glasses are also used for calibration, a more dependable method involves attachment of a calibration gage to the suction line of the metering pump. Bypass valves are required to shut off suction from the tank and to reroute the flow through the gage for direct measurement of flow. A system employing such gages is shown in Figure 4.2.
2. Three-way flushing valves operated manually or automatically by solenoid to allow the system lines to be flushed when settling type slurries such as caustic, lime or diatomaceous earth are being handled.
3. Flowmeters to allow back-up calibration or as a monitor for the ongoing process. Flowmeters coupled with annunciators allow feedback for process control and are usually installed as part of some downstream operation.
4. Additional safety relief valves to prevent system overpressure.
5. Steam jackets to permit heating or to maintain process fluid temperature.

6. Metering rate timers and counters to totalize pump pulses and to shut off flow when a preset limit has been reached.
7. Valves to automatically open and close feed to the pump or to reroute discharge from the pump back into the tank after certain periods of operation. Valves may be manually controlled or actuated by hydraulic, pneumatic, or electric power.

VIII. MAINTAINING THE SYSTEM

Although chemical feed systems require little maintenance during their lifespan, a few basic steps are necessary for optimum performance. For specific pump maintenance procedures, see Chapter 10, Maintenance Guidelines, or consult the pump manufacturer.

The following maintenance checklist applies to the remainder of the system:

1. Check painted surfaces for scratches and evidence of corrosion. After proper surface preparation, apply a suitable primer and finish coat.
2. Inspect piping and tubing for bent or nicked sections. Replace damaged portions.
3. Check pipe supports for cracks due to vibration and other mechanical loads. Initiate appropriate repairs.
4. Inspect base or skid drain holes for possible clogging.
5. Clean strainer screens and sludge traps to prevent line blockage.
6. Flush base or skid to remove spilled chemical residue.
7. Clean chemical basket inside tank to insure proper mixing during agitation cycle.
8. Clean and check level probes or other electrodes in tanks and piping.
9. Check agitator shaft and blade for buildups. Unbalance can lead to rapid bearing wear.
10. Clean gauge glasses to ensure proper visibility at all times.
11. Lubricate the stem and service packing on all manually operated valves within the system. Cycle in and out to ensure proper operation.
12. Check grounding connection on motors and ensure that fan intake is clear.
13. Cycle relief valves to ensure proper operation.
14. Remove and calibrate pressure gauges.
15. Throughly inspect panels to ensure seal integrity. Water penetration due to outdoor exposure can lead to severe consequences.

5

METERING PUMP CONTROL SYSTEMS

The preceding chapters provide an overview of the major types of metering pumps, chemical feed systems and accessories. This chapter discusses the control systems that are configured with this equipment.

Industrial processes frequently require the mixing or proportioning of two or more fluids in a specific volumetric ratio. At one time, this could be accomplished only as a batch process; a tank or vessel would be filled with a predetermined volume of liquid, and other chemicals would be measured and added manually. The many problems and inaccuracies that make this type of proportioning unacceptable are resolved by the metering pump. Today, the metering pump is used not only in simple batching and dosing systems, but also in open, closed and combined open/closed loop proportioning systems.

I. SIMPLE ADDITIVE SYSTEMS

Most batch-type processes are fairly stable; the same can be said for certain continuous process applications. In either case, frequent metering pump capacity changes are not necessary—the quantity of the chemical(s) to be added is fairly constant. Metering pump operation can therefore be timed manually or by an automatic timing control device to permit delivery of the required volume. In such applications, pump output is fine-tuned as required by manual capacity adjustment.

A simple additive system for a batch-type operation consists of a supply tank to store each metered chemical, a metering pump (or liquid end) for each supply tank equipped with a manual capacity controller, a large batch tank for mixing the chemicals, and, optionally, a timing device (see Figure 5.1). Such a system represents an improvement over manual addition of chemicals by eliminating human interface and simplifying the means for accurately transferring chemicals from storage to the mixing tank.

Simple additive systems are also used in certain continuous process applications in which conditions do not vary significantly. When metering sodium hypochlorite into a fairly constant water stream, for example, the pump output can be set to provide the proper ratio on a continuous basis. The use of a simple

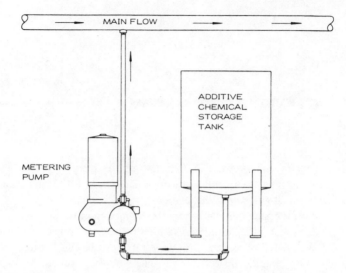

Figure 5.1. When blending conditions are relatively constant, a constant speed metering pump can be used to add a chemical in proportion to mainstream flow.

additive system in continuous process applications requires periodic downstream sampling to verify that the ratio of additive to main fluid falls within prespecified limits.

Because many mainstream flows are not constant, however, and precise ratios must be maintained, other types of systems that can automatically adjust dosage rates or respond to process feedback are often preferred.

II. OPEN LOOP SYSTEMS

When mainstream flows vary widely, the addition of a fixed volume will not maintain the specified ratio. Open loop control systems are normally used for automatically adjusting the pump output in response to these changes in flow. This is accomplished through use of a flowmeter in the main process line as shown in Figure 5.2. As it is usually not necessary to attain zero flow conditions in such systems, the instrument signal from the flowmeter is used to control a variable-speed motor (via an SCR or adjustable frequency AC controller) that serves as the pump driver. Pump output is adjusted by altering the stroking rate per minute (spm) in response to the flowmeter signal. This permits ratio (dosage) adjustment to be effected by way of the manual capacity control knob on the pump.

SCR (silicon-controlled rectifier) motor speed controls, have until recently, been preferred due to their higher efficiency and lower cost. Recent develop-

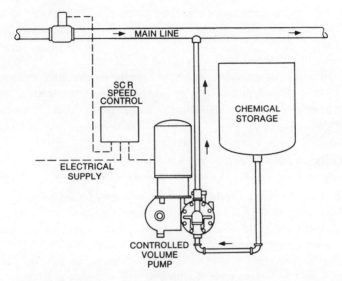

Figure 5.2. Open loop flow control is used to change rate of chemical injection in proportion to variations in main line flow.

ments in AC motor speed controls, notably adjustable frequency devices, have all but eliminated this advantage; thus selection is now based mainly on environmental and other factors. DC motors cannot be used in hazardous environments, for example.

If AC motor speed control is elected, the controller must be sized to handle the *maximum* torque requirements of the metering pump. Due to the sinusoidal performance characteristic of reciprocating pumps, and related inertial and friction effects, the start-up torque can approach 300% of running torque at maximum operating pressure. Because many AC controllers are rated for a 150% overload under start-up conditions, performance specifications must be carefully considered.

Turn-Down Ratios

The desired turn-down ratio of the metering pump is an important consideration in control selection. Most metering pumps offer 10:1 turn-down ratios, which means the output can be very accurately adjusted between 10 and 100% of rated flow. Below 10%, however, it is difficult to achieve accurate delivery without further refining the accuracy by combining effective stroke control with motor speed control. With the motor running at 1/6 of rated speed, for example, and the capacity adjustment set at 10% the actual pump output would be 0.167 x

10% of total pump capacity. Turn-down ratios can be so extended until the minimum effective stroking speed—15–20 spm for most metering pump designs— is reached.

When turn-down ratios in excess of 3:1 for motor speed controllers are desired, a tachometer should be added to the control circuitry. Feedback from the tachometer is used to fine-tune motor speed and eliminate synchronization problems caused by the varying torque characteristics inherent in reciprocating mechanisms.

III. CLOSED LOOP SYSTEMS

Closed loop systems (see Figure 5.3) are characterized by feedback or corrective action from the controller. Such controllers are quite sophisticated, and read the process variable after the addition of the chemical reagent by the pump. The accuracy of the pump and its control (pneumatic or electric) is relatively important to maintaining the process variable close to the setpoint, or desired process conditions, without considerable modulation by the process controller.

In closed loop systems, it is essential that the pump capacity be able to go to zero. It is more practical to attain zero output with the stroke control than by varying motor speed, as many variable-speed drives cannot be satisfactorily operated at zero speed. Capacity controls employed in closed loop systems are usually pneumatic or electric analog devices.

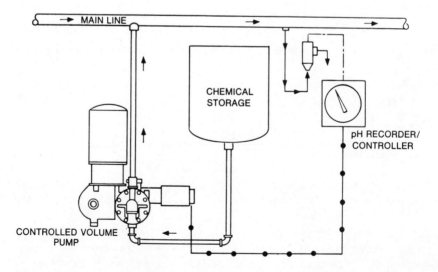

Figure 5.3. Closed loop flow control is used to change rate of chemical injection to maintain a process variable with precise limits.

Closed loop systems are normally used when the mainstream flow rate is relatively constant but a process variable, such as pH, changes. In an automatic pH control system, for example, pH is dependent on the reaction between the metered chemical and the raw liquid. The pH sensor is installed at least 20 diameters downstream of the mixing point to permit thorough mixing, and typically emits a 4 to 20 ma instrument or 3 to 15 psi pneumatic signal that is proportional to deviations from the pH control point. This signal actuates an electric or pneumatic capacity control unit on the pump, which increases or decreases the amount of control agent delivered.

IV. OPEN/CLOSED LOOP SYSTEMS

When both the process variable and flow rate change appreciably, a metering pump can be used as the final control element for two process variable controllers in a combination open/closed loop system.

An example would be a pH control system with widely varying flow rate. For automatic control, an upstream flowmeter and downstream sensor/pH controller are superimposed on one metering pump with variable motor speed and automatic capacity control. Pump speed is controlled by the variable-speed motor; stroke length is adjusted by the capacity control. Such a system can handle large sweeps in flow rate as well as changes in pH and can offer turndown ratios as great as 100:1. A typical open loop proportional feed system cascaded to a closed loop process variable control system is shown in Figure 5.4.

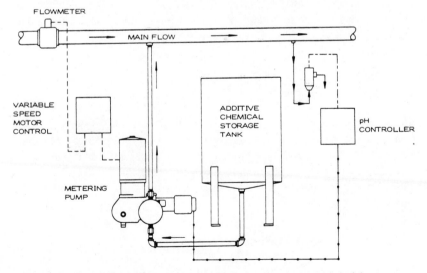

Figure 5.4. Open/closed loop control. Combined open and closed loop system using a metering pump as the final control instrument.

V. PNEUMATIC SYSTEMS

Pneumatic drives and controls are often utilized in extremely hazardous atmospheres. Pneumatic pumps (Figure 5.5) employ a direct-coupled air cylinder drive to impart back-and-forth action to the pump plunger; liquid ends may be of the packed plunger, disc diaphragm or tubular diaphragm type. The type of liquid end selected depends on the nature of the fluid to be metered (see Chapter 3).

The direct-coupled air cylinder drive provides added control versatility and eliminates the need to convert rotary electrical motion into reciprocating plunger motion; thus, pneumatic pumps are well suited for batching and dosing operations in which intermitent injection or variable delivery rates are required. They may also be used in continuous proportioning applications in which chemical injection must be varied to meet the mainstream flow.

A typical pneumatic system consisting of two metering pumps, digital pacing devices and totaling counters is shown in Figure 5.6. These electrical controllers are used singly or in combination to tailor the system to the requirements of the process; in addition, capacity can be adjusted manually via the capacity control handknob, and independent adjustment of plunger traverse speed is available for pumping high viscosity fluids and for improved suction lift capability.

A. STROKE PACER

The stroke pacer is similar to the variable-speed motor in that it permits pump stroke rate to be adjusted from 1 to 150 (or more) strokes per minute; thus, delivery rate can be precisely controlled. Upon actuation, the pump operates at the preset stroke rate until it is commanded to stop.

B. ANALOG RATIO PACER

This controller allows the pneumatic pump to respond instantaneously to analog signals from process control instrumentation and adjusts its discharge rate accordingly. This is accomplished through conversion of a milliamp input signal into a pulse rate output of 1 to 150 (or more) strokes per minute. For most pneumatic pumps, the ratio of pulse rate output to analog signal input is adjustable over a range of 100 to 20%.

A. Pneumatic Proportioning Systems

Instantaneous proportionality is easily and economically obtained with pneumatic proportional feed systems. In a typical system, a flowmeter is mounted in

Figure 5.5. Pneumatic metering pump used to accurately deliver adhesive used in a high-volume manufacturing process.

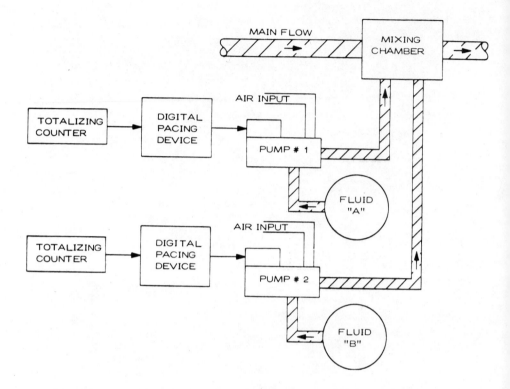

Figure 5.6. Pneumatically-operated pumping system. Several pneumatically operated pumps are used in the multiple chemical blending system shown. Each digital pacing device can be used to independently vary additive flow rates, and a totalizing counter can be employed to determine cumulative flow.

the main flow line. Several types of meters may be used, including nutating disc and rotary vane meters of the positive displacement type and shunt-current and propellor meters of the velocity measuring type. Although the sensing element in each meter is different, each translates the measured variable (either a fixed volume or a velocity that is proportional to volume) into a rotary output shaft movement. The speed or shaft rotation is directly proportional to flow.

The shaft is coupled with a four-way rotary air pilot valve which sends a sustained air impulse (e.g., twice each revolution) to a master four-way valve. A piston in this master air valve receives the impulses from the meter and reciprocates with each pulse. As the piston reciprocates, it alternately admits air to one side and then the other of the air cylinder on the pump. The pump, which controls the flow of chemical, completes a suction and discharge stroke with each cycle of the master four-way valve. The quantity of chemical delivered by the pump is therefore directly proportional to the main flow measured by the meter.

For example, the shaft (on the rotary air valve mounted on the main line meter) could be geared to make one complete revolution for each two gallons of flow in the main line. At a flow rate of fifty gallons per minute, the shaft would complete twenty-five revolutions, thereby stroking the pump fifty times per minute. If main line flow dropped to twenty-five gallons per minute, the pump would stroke twenty-five times.

VI. MULTIFLUID METERING

Multifluid metering equipment that automatically delivers a predetermined ratio as the process flow rate varies over the full range of capacity includes the four basic functions shown in Figure 5.7. As previously mentioned, the volume of the base fluid (usually the greater flow rate and thus called the "mainstream") is measured. A signal that is proportional to the flow rate is developed by the meter. This signal is conditioned to match the span and range requirements of the control element. Finally, the control adjusts the stroke length or the stroke speed of the metering pump to provide the required delivery.

When a closed loop control configuration is necessary, the basic functions are as shown in Figure 5.8. The closed loop control includes a process monitor in the mainstream after the chemical addition, with a signal conditioning interface and delivery control element for one or more of the metering pumps. The process monitors include all the various instruments for measurement of conditions such as pH, conductivity, color, heat of reaction, etc.

The equipment and systems used can be classified by one of the following general headings:

1. Fluid meter *driven* metering pumps.
2. Fluid meter *actuated* metering pumps for intermittent delivery.
3. Fluid meter *paced* metering pumps for continuous delivery.
4. *Total* metering for optimum flow ratio accuracy.

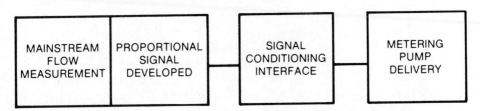

Figure 5.7. Four basic functions of open loop multifluid metering equipment.

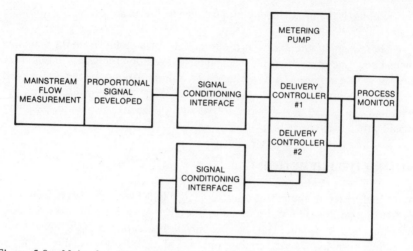

Figure 5.8. Major functions of closed loop multifluid metering equipment.

A. Fluid Meter Driven Metering Pumps

The simplest way to accomplish the required functions for open loop/constant ratio proportioning is to use a fluid metering device that develops reciprocating motion as an output signal. This motion is used to drive the metering pump. The advantages of this type of system are:

1. Ease of installation and operation; no external power required.
2. Low initial cost.
3. Positive displacement fluid motor insures consistent proportioning even at low flow rates.
4. With constant operating conditions a proportioning ratio of ±5% can be maintained over the mainstream capacity range.

The disadvantages and/or limitations of this equipment are:

1. The adjustment and confirmation of a specified proportioning ratio is difficult due to the coarse and indirect means for adjusting the chemical volume.
2. The double-acting piston and directional control valves in the fluid motor are subject to rapid wear if foreign particles are present in the mainstream.
3. Only dual fluid proportioning is available. Additional units would be required for each additional chemical.

Similar proportioning devices are available which use a nutating disç type fluid meter in place of the double acting fluid motor. The chemical pump plunger is driven by an eccentric cam mounted on the rotating output shaft of the meter. The limited torque to drive the plunger restricts the pump discharge to near atmospheric pressure and reduces the reliability of this configuration.

Typical applications for fluid meter driven pump proportioning equipment are:

1. Proportioning water and coolant fluids for metal cutting machines.
2. Adding chemicals to low-pressure water systems.
3. Adding chemicals to cooling tower make-up water.
4. Proportioning liquid fertilizers to water sprinkling systems.

B. Fluid Meter Actuated Metering Pumps for Intermittent Delivery

Most applications for automatic constant ratio proportioning with a variable mainstream flow fall under this classification. Many types of flowmeters, signal conditioners and metering pumps are used. Each application is characterized by the intermittent, single-stroke or start-stop operation of the pump to deliver a controlled volume of chemical for a fixed quantity of mainstream flow through the meter.

Automatic constant ratio proportioning requirements are frequently approached from a systems concept. A fluid meter with a known output signal characteristic is combined with a signal conditioning device to control the pump delivery. The components used for these systems are shown in Table 5.1, which includes both positive displacement meters and flowmeters.

The positive displacement meters of the nutating disc, oscillating cylinder and oval gear types are used to provide a digital type electric pulse signal for a specific flow through the meter. This signal pulse is used either to actuate one or more single-stroke pump mechanisms or to actuate a timer which provides for delivery from one or more pumps for a predetermined period.

The fixed- and variable-area orifice flowmeters and venturi meters normally develop an analog signal based on differential pressure or relative position of the metering device. The analog signal is conditioned to a timed electric signal with the duration proportional to the measured mainstream flow. The intermittent timed signal then actuates a solenoid bypass valve that determines the stream from a continuously operating metering pump.

The operating characteristics and advantages of these systems include:

1. Mainstream flow rates up to 500 GPM can be measured with positive displacement meters. Higher flow rates can be measured using fixed

TABLE 5.1

Fluid Meter Actuated Metering Pumps for Intermittent Delivery

Mainstream Flow Measurement	Proportional Output Signal Condition	Metering Pump Configuration
1. Positive displacement meters, including switch closure mechanism on output shaft (nutating disc oscillating cylinder oval gear types)	Electric signal pulse each specified increment of mainstream volume	(a) Signal actuates single-stroke hydraulic, pneumatic or electric drive driven pump (b) Signal actuates timer which turns pump on for a pre-set period (c) Signal actuates timer, which energizes bypass solenoid valve to control delivery of chemical into system (Pump operating continuously)
	Hydraulic or pneumatic directional control valves operated each specified increment of mainstream volume	Actuates single stroke hydraulic or pneumatic cylinder driven pump
2. Orifice meters, venturi meters	Differential pressure proportional to flow is converted to an electrical signal with a timed duration proportioned to mainstream flow	Timed electrical signal ener-control delivery of chemical into mainstream (Pump operating continuously)
3. Variable-area orifice	Vertical position of the weight plug is converted to an electrical signal with a timed duration proportional to mainstream flow	

Source: Milton Roy Company

orifice meters installed in a bypass configuration can also be used for the higher flows.

2. Proportioning ratios of 100:1 (mainstream/chemical by volume) and greater can be obtained with one or more chemicals.
3. A proportioning ratio accuracy of ±5% can be obtained automatically over a 10:1 mainstream flow range.
4. Maximum operating pressures are determined by the type of fluid meter used. (Normally 125 psi for positive displacement meters and higher pressures for the orifice, venturi and variable-area orifice flowmeters).

The operating limitations and disadvantages are:

1. The intermittent injection of chemicals into the mainstream results in extreme variations in downstream chemical concentrations. This requires that a large-capacity mix vessel or storage tank be included in the process line before the point of application.
2. The positive displacement meters should only be used in clean, clear streams which do not contain abrasive particles or foreign materials. (The venturi meters can be used for dirty or unfiltered streams.)
3. Continuous operation of rapid-cycling, start-stop and single-stroke mechanisms often reduces the service reliability and accuracies of the system after an extended period.

Typical applications using this automatic constant ratio proportioning with intermittent chemical injection are:

1. Adding de-icer fluid and other modifying additives to petroleum fuels at the time of transfer to the tank truck.
2. Injection of the correct formulation of one or more liquid fertilizers into large commercial irrigation sprinkling systems.
3. Adding water treatment chemicals into pressurized circulating systems and cooling tower makeup water.
4. Multifluid proportioning for industrial and chemical processes.

C. Fluid Meter Paced Metering Pumps for Continuous Delivery

Operating systems have demonstrated the advantages of using equipment that automatically maintains the specified volume ratios for multifluid formulations. However, continued development of this concept is required to improve the range of operation and ratio variation tolerance and to eliminate the problems of both meter driven pumps and meter actuated pump systems. Optimum performance of controlled volume pumps can be realized by fluid meter paced operation.

The flowmeter signal is used to vary the effective plunger stroke length or the plunger stroking speed so that the delivered chemical volume is directly proportional to the measured mainstream. Pump capacity controls are available that automatically adjust the effective plunger stroke length in response to an analog electric signal (resistance, voltage or current) or pneumatic signals. These controls can be paced automatically from a remote analog signal (either resistance, voltage, or current). Application experience with systems using meter paced pumps with continuous delivery has demonstrated improvements in:

1. Proportioning accuracy.
2. Uniformity of injected chemical dispersion throughout the mainstream.
3. Overall service reliability.

Table 5.2 lists the various types of meters, output signals and methods of adjusting pump delivery.

For maximum range of operating capacity and proportioning accuracy, it is recommended that the mainstream flowmeter be used to control pump stroking speed via a variable-speed motor. The stroke length adjustment is used to accurately set the required capacity for the individual pumps. This adjustment is also available for closed loop control of the delivery on any individual pump.

Typical performance curves for the components of a mainstream flowmeter paced variable-speed pump motor are shown in Figure 5.9. The mainstream flow versus signal voltage is a linear curve through the zero intercepts, but the motor speed versus signal curve shows that a definite voltage signal level is required (threshold voltage) above zero to obtain a minimum motor speed.

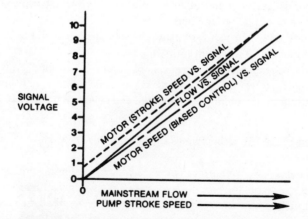

Figure 5.9. Typical performance curve for mainstream flowmeter paced variable speed pump drive motor.

TABLE 5.2

Fluid Meter Paced Metering Pumps for Continuous Delivery

Mainstream Flow Measurement	Proportional Output Signal Condition	Metering Pump Delivery Adjustment
Positive displacement nutating disk meter with direct driven DC generator (up to 300 gpm)	DC voltage signal	(a) Electronic SCR variable speed controls for pump drive motor (b) Electric stroke length adjustment
Flow transducer— Variable reluctance (up to 500 gpm)	DC voltage signal or current signal	
Turbine Flowmeter	Digital pulse rate is converted to proportional analog voltage or current signal	
Variable-area Orifice Flowmeter	Position of weight plug is mechanically linked to potentiometer for proportional resistance signal	
	Position of weight plug is converted to proportional pneumatic signal	Pneumatic stroke length adjustment
Fixed orifice and venturi Flowmeter	Differential pressure signal converted to proportional pneumatic signal	
	Differential pressure signal converted to proportional resistance, voltage or current signal	(a) Electronic SCR variable-speed controls for pump drive motor (b) Electric stroke length adjustment

Source: Milton Roy Company

Since pump delivery is a direct function of drive motor speed, the difference in the zero intercepts for the two curves represents a change in proportioning ratio with a change in mainstream flow. Until recently, this limited the constant ratio proportioning capabilities of this system to a 2:1 flow range. However, SCR control circuits can be biased with an independent voltage signal equivalent to the threshold voltage so that the two curves pass thru the zero intercepts. This modification to the SCR control circuit significantly expands the limits for the constant ratio proportioning to a 15:1 range of mainstream flow rate with a ratio accuracy of ±3%.

When it is necessary to add more than one chemical, metering pumps can either be multiplexed together to use a common variable speed drive motor or the signal can be applied to several variable speed controls for the drive motors on individual pumps. As shown in Table 5.2, this meter-paced variable-speed proportioning is available over the entire range of mainstream flows, through use of many different types of flowmeters.

D. Total Metering for Optimum Flow Ratio Accuracy

The preceding systems have been based on measuring one stream flow and automatically adjusting the delivery of the metering pumps to maintain the specified proportioning ratio. When a ratio accuracy of better than ±3% is required, it can be obtained by using a metering pump to control the delivery of each fluid. For this flow ratio control system, an electronic variable-speed motor is used to determine the stroking speeds for all the multiplexed pumps so that the total delivery can be varied over a 10:1 capacity range. The proportioned ratio is set by adjusting the effective plunger stroke length for each individual pump. This system utilizes the repetitive accuracy capabilities of the metering pumps for proportioning each fluid so that a constant flow ratio accuracy of ±1% can be obtained over a 10:1 range of total capacities.

6

INSTALLATION

The first step in installing any metering pump is to study the manufacturer's instructions carefully. Instructions on lubrication of motors, gears and other moving parts are especially important (see Chapter 10, Maintenance Guidelines). Only the best lubricants should be used. Wherever special types of lubricants are recommended for specific purposes, there should be no substitution.

I. LOCATION AND MOUNTING

The pump should be located so that its suction connection is well below the liquid level of the source tank. This will provide full flooded suction. Although metering pumps operate with suction lifts of up to 20 ft of water, they may perform erratically if even slight leakage occurs in suction piping or at the pump stuffing box. In addition, the pump should be located as close as possible to the source tank to minimize pressure loss in the suction piping.

   The pump should be firmly bolted to a solid foundation, preferably a concrete pad. The pad should be large enough to leave plenty of room around the pump for easy service access. It should also be high enough to protect the pump drive mechanism from flooding should the area be washed down.

II. PIPING

Many in-field problems are caused by undersized piping. Piping must be large enough to handle peak instantaneous flow, which is considerably greater than the rated pump capacity. Due to the reciprocating motion of the plunger, the flow in motor-operated pumps follows a sine curve. Peak instantaneous flow is approximately 3.14 (or π) times the average flow.

   To check sizing, unhook the suction tubing and measure the free flow of liquid from the pipe. If the flow is approximately 3.14 times the rated capacity of the pump, piping size is usually adequate, provided that piping is relatively short in length (10 ft or less).

Use piping that is resistant to corrosion by the process liquid and match inside diameters perfectly between connections. Select a material that will avoid galvanic corrosion at pump connection points. Burrs and sharp edges should be removed and weld splatter cleared from the lines. All lines should be blown or flushed before final connections to the pump are made. If hot or cold liquids will be pumped, allowance should be made for expansion and contraction of the piping.

Metal or plastic tubing perform better than standard pipe in suction lines. Tubing has a smooth inner surface and can be formed into long, sweeping bends to minimize frictional flow losses.

Suction and discharge lines should be as short and straight as possible, and unnecessary elbows should be avoided. Where possible, use 45° or long-sweep 90° fittings. Piping should be sloped to eliminate vapor pockets—vapor in the liquid end can cause inaccurate pump delivery. Additionally, piping should be supported to avoid strain on the pump and should never be laterally loaded or "sprung" when connections are made. The number of fittings should be kept to a minimum in order to reduce pressure losses. This is especially important if suction lift is required or if the liquid being pumped is close to its boiling point.

A. Suction Piping

Suction piping must remain absolutely airtight. After installation, piping should be tested for leaks. In large-sized pumps operating at higher speeds, air in the suction system can result in a form of "water hammer."

Water hammer results when the pump chamber fills partially with liquid and partially with vapor on the suction stroke. On the discharge stroke, the plunger moves forward, collapsing the vapor bubble. As the plunger continues forward at high velocity, it "strikes" the liquid in the chamber. The resulting shock produced vibration in the entire system and can cause serious damage to both piping and pump.

When pumping from a tank, the suction pipe should enter several inches above the tank bottom. This leaves a sludge-settling zone. Such care is especially important with low-capacity pumps (30 gph or less) because their small check valves are susceptible to fouling by undesirable particles and sludge.

Adequate net positive suction head (NPSH) must be maintained to ensure proper operation. NPSH is the fluid pressure above the vapor pressure of the liquid being pumped that is available to accelerate and move process liquid into the pump. (Appendix C details calculation of NPSH.) The potential for vapor formation, or flashing during pump suction, increases in proportion to pump speed, fluid viscosity and density, plunger size, length of piping and lift height. Insufficient suction head pressure may result in pump starvation or cavitation in the liquid end, which destroys the metering action of the pump.

Two major factors can adversely affect the metering function:

1. *Flow losses* relate to the pressure required to ensure the liquid flows continually in the system during the peak flow demand portion of the pump stroke.
2. *Acceleration losses* refer to the pressure required to cause the liquid to move at the beginning of each plunger stroke.

Calculating NPSH determines which of these two factors is dominant in a given application, and therefore the limiting consideration. Generally, for water-like liquids with viscosities below 50 cP, acceleration losses are highest. Practical means for improving NPSH include:

1. Increasing inlet pipe diameter.
2. Decreasing pump speed while increasing plunger diameter to maintain flow rates.
3. Increasing head pressure by raising the level of the supply tank relative to the pump inlet.
4. For high-vapor-pressure fluids, pressurizing the supply tank.

The plunger displaces a definite volume of fluid during each cycle. Assuming it is properly fitted to the displacement chamber, the volume of fluid displaced is equal to the volume of a cylinder having the same diameter as the plunger and length corresponding to stroking distance. The actual amount of liquid pumped may vary from this theoretical value for the following reasons:

1. Fall-back loss or excessive time associated with check valve operation.
2. High discharge pressure (fluid compressibility).
3. Entrapped gases.
4. Pump head elasticitiy.
5. Leakage losses through defective piping and seals.

The actual amount of liquid displaced by a properly operating pump is about 95% of the theoretical displacement, assuming the liquid being handled is not compressible. A simplified equation for determining volumetric efficiency is:

$$\frac{\text{Volumetric}}{\text{Efficiency}} = \frac{\text{Delivered Volume}}{\text{Swept Volume}}$$

If the pump is not operating at rated capacity, its volumetric efficiency may be calculated by modifying the equation slightly:

$$\frac{\text{Volumetric}}{\text{Efficiency}} = \frac{\text{Actual Flow}}{\text{Theoretical Delivery} \times \text{Percent of}}$$
$$\text{@ Maximum Stroke} \quad \text{Stroke}/100$$

The term "compression ratio" does not normally apply to pumps handling non-compressible fluids; however, the compression ratio of a metering pump becomes important under the following conditions:

1. The ability of a pump to recover from vapor lock.
2. The ability of a pump to lift liquids and prime against process pressure.
3. Accuracy and repeatability when pumping liquids of different compressibilities at high pressures.

The compression ratio of a metering pump is defined as the volume of the displacement chamber when the plunger is at the extreme end of the suction stroke divided by the volume of the chamber when the piston is at the extreme end of the discharge stroke, or unswept volume.

High compression ratios are achieved by designing the liquid end so that the piston swept volume approaches the total volume of the fluid displacement chamber. For example, a pump with a 1200 cc fluid displacement chamber and 1100 cc swept volume will have a compression ratio of $\dfrac{1200}{1200-1100}$ or 12:1. A pump with a 12:1 compression ratio will recover from total vapor lock and continue to deliver fluids up to a discharge pressure of approximately 175 psia.

Given pump compression ratio (CR), suction pressure (Ps), and discharge pressure (Pd), the percentage of vapor lock (VL%) from which the pump can recover is expressed as follows:

$$\left(\frac{1 - \dfrac{1}{CR}}{1 - \dfrac{Ps}{Pd}} \right) \times 100\%$$

Note that pumps with a CR greater than 10:1 easily recover from vapor lock, while pumps with CRs less than 2:1 have a difficult time clearing the smallest bubbles.

Assume that the flow capacity of a 12:1 compression ratio pump may be adjusted by changing stroke length. When capacity is decreased to 25% of maximum stroke, the compression ratio becomes 1.30:1. Thus, linearity over the entire stroke range is related not only to check valve efficiency but also to fluid compressibility and entrapped gases.

B. Discharge Piping

Discharge piping should be large enough to prevent excessive pressure losses during the discharge stroke of the pump. Maximum pressure at the discharge fitting on the liquid end must be kept at or below the maximum pressure rating shown on the pump data plate. Piping should be arranged to provide at least 30–50 psi positive pressure differential on the discharge side of the pump. The pump will not deliver a controlled flow unless this differential exists—the discharge line pressure must be at least 30 psi greater than the suction-line pressure. If necessary, artificial discharge pressure can be created by installing a vented riser or a back pressure valve (see Chapter 2, Metering Pump Accessories).

III. MISCELLANEOUS START-UP CONSIDERATIONS

A. Pump Drive Motor

Connect the correct power supply according to motor nameplate. Be sure motor shaft rotation is in the same direction as that indicated on the drive side flange of the pump.

B. Stuffing Box

When a packed plunger pump will be used to pump suspended solids and abrasives, internal flushing of the packing is necessary. Although the stuffing box is designed to handle most clear, free-flowing liquids, those with suspended solids and abrasives (e.g., certain slurry and phosphate solutions) tend to preceiptate in the packing, causing abnormal wear on packing and plunger. An internal flushing connection used with a V- or Chevron-type packing (see Chapter 10, Maintenance Guidelines) counteracts this tendency and increases packing and plunger life.

To connect for internal flushing, remove the stuffing box grease fitting and connect the stuffing box to a source of water or other compatible liquid at 25 to 50 psig above suction pressure. Since only a few drops per minute are necessary, small-diameter tubing will suffice. Install a 1/8-in. or 1/4-in. NPT stainless steel aircraft hydraulic system type check valve on the flush line right next to the stuffing box connection. This keeps the process liquid from backing up through the flush line should the packing wear or fail. A 1/8-in. or 1/4-in. needle valve should be included for controlling the flushing liquid flow rate.

When hazardous or undesirable fluids are being handled by a packed plunger pump, through flushing of the stuffing box is a must. Connections can be provided for through flushing by drilling and tapping the stuffing box during manufacture. Through flushing carries the undesirable packing leakage away from the stuffing box to a drain or other suitable disposal point.

C. Drains

Be sure drains are located convenient to the pump for easy disposal of any leakage.

D. Packing

Packing should be broken in before the actual pumping load is applied to a packed plunger pump. This is done by releasing the packing, pulling it up just a little more than "fingertight" and injecting packing lubricant into the lubrication fitting that is provided. The packing gland should then be pulled up tightly to "set" the packing in the packing box. Keep the packing loaded for several minutes to allow the packing to flow into position in the stuffing box. Release the packing and draw up the gland a little more than fingertight. Operate the pump dry (i.e., without pumping load) for about one-half hour, occasionally applying a few drops of light oil to the plunger where it enters the stuffing box.

It is a good practice, as well, to start operation at zero stroke and then gradually increase stroke length if the pump has been in storage for any length of time. Plasticizers in the packing can leach out and act as an adhesive between packing and plunger. Gradual start-up will usually break the bond without any packing or pump damage.

The pump should then be placed in operation and adjusted for proper feed. If there is leakage at the packing gland, tighten the packing *very gradually* over a period of several hours.

The first 24 hours of operation are very critical as far as packing life is concerned. If the packing is pulled up too tightly, it will overheat and disintegrate rapidly. Excessively tight packing may score the plunger, necessitating its replacement. If packing is very carefully adjusted during the first 24 hours of operation, it will develop a good running surface, and packing life will be greatly extended.

Packing should be lubricated regularly according to the manufacturer's instructions. Lubrication periods may range from daily to more lengthy intervals, depending upon the type of liquid pumped, the temperature of liquid, the type of packing lubricant and other variables (see Chapter 10, Maintenance Guidelines).

Figure 6.1. Typical micrometer handknob capacity control set at 100% of pump capacity.

IV. FINAL CHECK AFTER INSTALLATION

Before using the pump, set the stroke to zero and operate the pump for 15 minutes. This will thoroughly distribute the gear oil throughout the pump gear box section. After the pump has been running for 15 minutes at zero stroke, gradually increase the stroke setting, first to 25%, then to 50% and finally to 100% (Figure 6.1), running the pump for 10 minutes at each setting.

After a new pump has been running for several hours, relubricate it thoroughly and check motor temperature and packing adjustment. In a warm environment such as a boiler room, the motor shell temperature may run as high as 190°F or so, which is normal for a small motor and reducer. This will pose no problem to the pump or motor provided the electrical overload device is correct for the motor, adequate ventilation is provided, and the reducer is correctly lubricated.

V. CALIBRATION

Run the pump for at least 24 hours before checking calibration. The calibration test should be run at 10%, 50%, 75% and 100% of full stroke. To minimize error, runs should at least encompass 100 pump strokes. Plotting the resultant data on

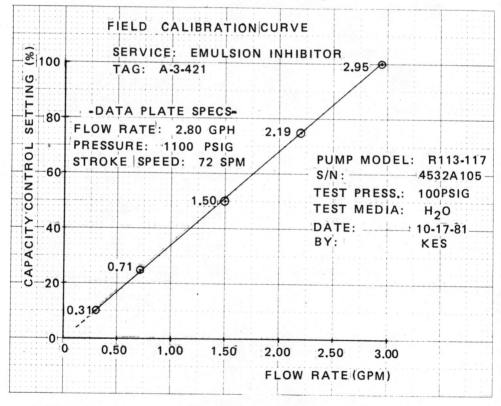

Figure 6.2.

graph paper provides a pump calibration curve (Figure 6.2). Since feed rate is proportional to stroke length, stroke settings for various feed rates can be determined easily by using this curve.

VI. WATER TESTING THE PUMP

Before pumping any chemical solution, test the system using water (when possible). Testing the system using water as the process fluid can help the user avoid hazards. The source tank should be filled with water to a point above the pump suction connection. After proper system operation is verified, close off the discharge shut-off valve to force the discharge through the relief valve, permitting its operation to be checked as well. Be sure to blow all water from the system at the conclusion of the test. When pumping concentrated sulfuric acid, it is essential that all piping and the pump chamber are completely free of moisture and thoroughly dry.

7

METERING PUMP APPLICATIONS

This chapter provides an overview of typical metering pump applications. Due to the myriad uses that can be found, it is not intended to deal with the minutiae of each process. Rather, those areas in which metering pumps are likely to interface are detailed. Metering pump applications by industry and a summary description of applications are capsulized in Tables 7.1 and 7.2, respectively. Due to its complexity, wastewater treatment in major industries is discussed separately in Chapter 8, Potable Water and Wastewater treatment.

Much of the information concerning water treatment chemistry and practices described in Chapters 7 and 8 is provided courtesy of Betz Laboratories, Inc. Further detail can be obtained from their excellent reference work [*Betz Handbook of Industrial Water Conditioning*, Betz Laboratories, Inc., Trevose, PA, 7th ed. (1976)].

I. CHEMICAL/PETROCHEMICAL PROCESSING

Major metering pump applications in the chemical/petrochemical processing industries include:

cooling tower water treatment;   process liquids;
internal boiler water treatment;   fuel additives;
external boiler feedwater treatment;   system flushing;
demineralizer regeneration;   process water treatment;
process additives;   wastewater treatment.

A. Cooling Tower Water Treatment

Cooling towers cool water by evaporation, lowering the temperature of water being returned from process cooling operations (Figure 7.1). In the chemical/ petrochemical processing industries, constant water temperature is necessary in order to produce a consistently high-quality product.

TABLE 7.1

Major Metering Pump Applications by Industry

Type of Application	Industry					
	Electric Power Generation	Chemical & Petrochem. Processing	Petroleum Refining Nat. Gas Processing	Water & Waste Treatment	Woodpulp Paper Processing	Food & Beverage Processing
Cooling Tower Water Treatment	●	●	●		●	●
Internal Boiler Water Treatment	●	●	●		●	●
External Boiler Feedwater Treatment	●	●			●	●

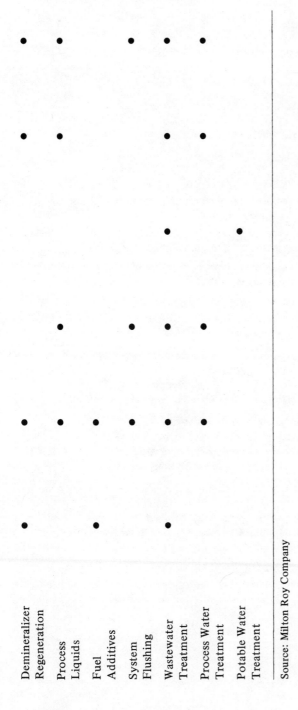

Source: Milton Roy Company

TABLE 7.2

Summary of Metering Pump Applications

Application	Explanation
Cooling Tower Water Treatment	Addition of chemicals to water in cooling towers to prevent or inhibit corrosion and fouling of metal heat-exchange surfaces.
Internal Boiler Water Treatment	Addition of chemicals directly into the high-pressure side of a boiler, as feedwater is either entering or inside the boiler. Chemicals are added to feedwater used in steam generation, to control scale deposits and corrosion.
External Boiler Feedwater Treatment	Addition of chemicals to feedwater supply in the low-pressure side of a boiler, as feedwater is being recirculated or discharged. This includes demineralizer regeneration
Demineralizer Regeneration	Injecting acid and alkaline solutions into demineralizer to flush accumulated salts from cationic, anionic, and mixed resin beds.
Process Additives	Injection of chemical additives directly into a process liquid or gas. This may be done to control corrosion and scale deposits, reduce foaming, initiate desirable chemical reactions or correct pH imbalance.
Process Liquids	Pumping liquids, often as part of the direct process stream, at precisely-controlled rates of flow through the various processing operations. This may be either in batches or as a continuous stream.
Fuel Additives	Addition of chemicals directly to fuel oil prior to burning. Fuel additives are injected in very small quantities, acting as a catalyst to improve combustion, minimize corrosion and deposits and improve boiler efficiency.
System Flushing	Injecting liquids used to help flush or clean out a process vessel or stream, to aid in preventing contamination, scale deposits or corrosion of process system elements.
Wastewater Treatment	Addition of chemicals to industrial and municipal wastewater to meet environmental requirements before discharge into drains, sewers, rivers and streams.

TABLE 7.2 *(continued)*

Application	Explanation
Process Water Treatment	Addition of chemicals directly into water used in a manufacturing process.
Potable Water Treatment	Addition of chemicals or other substances to purify water used for human consumption.

Source: Milton Roy Company

As water evaporates in a cooling tower, its dissolved solids become more and more concentrated. Unless chemically treated, these dissolved solids are gradually deposited as scale on metal surfaces, decreasing the inside diameter of tubes and pipes. This reduces their heat transfer efficiency and the volume of water flowing through the heat exchangers.

Blowdown, which replaces concentrated water with fresh make-up water, is one method of controlling scale deposits. A more effective method is to combine blowdown with the addition of one or more of the following chemicals to inhibit scale formation, minimize corrosion, control pH of the recirculating water and prevent biological growth:

biocides;
acids – H_2SO_4;
corrosion inhibitors;
dispersants.

Since these chemicals must be precisely controlled in concentration and amount, metering pumps are used to deliver them in carefully measured quantities and at specific rates. Typical injection points are shown in Figure 7.2.

It is desirable to feed treatment chemicals neat (undiluted) to the cooling tower water sump. The method is simple and permits a quick check of chemical pump operation. Acid can be fed with makeup water to a mix fitting or dilution trough. Feed systems for handling corrosion inhibitors and deposit control agents usually are equipped with a diaphragm-type pump. Non-oxidizing biocides are typically added over a brief interval (15 to 30 minutes) during each 24 hour period, necessitating a timer-operated pump.

When the cooling tower is remote or the installation is complex, it may be necessary to feed treatment chemicals to some point in the cooling water system other than the cooling tower sump. Inhibitors and deposit control agents may be injected anywhere in the recirculating system except immediately upstream of a sampling connection. Sulfuric acid sometimes may also be injected into the

Figure 7.1. Cooling towers are used in a variety of industries, including electric power generation (above).

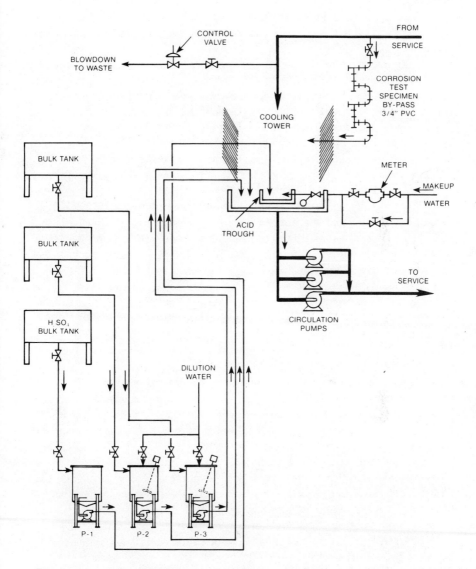

Figure 7.2. Chemical feed to a cooling tower system. [Reprinted from *Betz Handbook of Industrial Water Conditioning*, Betz Laboratories, Inc., Trevose, PA, 7th ed. (1976).]

recirculating system if certain limitations and precautions are observed to prevent local corrosion.

Acid is an important adjunct in the treatment of many waters used in cooling tower systems. It is used to restrict pH and reduce the total alkalinity concentration.

Acid treatment helps prevent the formation of calcium carbonate scale, deposition of calcium phosphate, and delignification of tower wood. pH control remains a critical element in many cooling tower water treatment programs even with the use of dispersants and other deposit control chemicals.

In cooling tower applications, concentrated sulfuric acid is normally injected at the sump or, occasionally, into a hot return. In the former case, the acid is diluted with make-up water in a static in-line mixer or an acid trough. The trough is located so as to discharge into the cooling tower basin at a point that allows thorough mixing prior to circulation.

When 66 degree Baume sulfuric acid is injected into a water line, it is necessary to ensure that it cannot be injected when water is not flowing in the mix zone. Corrosion-resistant piping should be used within eight diameters downstream from the injection point. Remote check valves should be installed on the acid inlet piping to preclude backflow of water into the acid feed system.

Because 66 degree Baume sulfuric acid attacks steel very slowly, steel is normally used for storage and day tanks. The tank must be closed and vented, preferably through a dryer. Unplasticized PVC (Type I or II) is satisfactory for piping and tanks for all concentrations of acid up to 93 percent (66 degree Baume), but precautions must be taken against physical damage. Schedule 80 piping with solvent connections should be used.

Dilute sulfuric acid and water solutions of powdered materials such as sulfamic acid or sodium bisulfate can never be handled in steel tanks, piping or pumps. Either 304 or 316 stainless steel, unplasticized PVC (vinyl) or rigid PVC are satisfactory. Although 93 percent sulfuric acid also may be handled in polyethylene tubing, it is very important to protect the tubing from physical damage and high temperatures. Polyethylene tanks are not recommended for concentrated sulfuric acid. They are satisfactory for dilute acid, but in making the dilution care must be exercised to dissipate the heat and prevent softening.

Proper design is important when installing a line for feeding of concentrated sulfuric acid to a cooling tower system where acid feed is controlled on/off by pH. The line must be installed so that slow filling and draining do not occur. Otherwise, when pH control calls for acid feed, a lag will occur as the line fills. After the acid begins to flow in the system, the pH control will stop the pump. However, as the acid line drains, the pH in the system will continue to drop, possibly to a seriously low value.

All horizontal sections of the acid line should flow upward in the direction of the flow. Installing a vented, elevated loop at the discharge end of the line, higher than the pump, assures a continuously filled line.

For cooling tower requirements, concentrated acid feed lines usually should be no larger than ½-inch schedule 80 unplasticized PVC pipe (Type I or II). Plasticized PVC (vinyl) will harden and is not recommended; 304 or 316 stainless tubing or pipe is excellent. Steel or black iron pipe is satisfactory, but when used in the suction of a pump not flooded, loss of prime can occur due to gradual hydrogen generation.

Dilute sulfuric acid or solutions of acidic powders, such as sulfamic acid or sodium bisulfate, may be handled in plasticized or unplasticized PVC, polyethylene, or stainless steel. Steel or black iron cannot be used with dilute acids.

Materials of construction in acid pumps usually will be unplasticized PVC, Teflon, alloy 20 or 316 stainless. Blue African asbestos and Teflon are suitable for gaskets and packing. Ceramic, glass or alloy 20 is used for ball checks.

B. Internal Boiler Water Treatment

Boilers use heat supplied by fuel to convert feedwater into steam under pressure. In the chemical/petrochemical processing industries, this pressurized steam is then used to initiate chemical reactions at high temperatures and pressures.

Corrosive attack on steel and common copper alloys is caused by the presence of aqueous acid solutions with a pH of 5 or lower at ambient temperature. Corrosion rates accelerate as either temperature or acidity is increased. Weak acids—such as aqueous hydrogen sulfide and hydrogen cyanide—are more corrosive than the pH of their solutions otherwise indicates. Under certain conditions, these acids are corrosive in the alkaline range. It should be noted that exceptions to the preceding statements can occur, however.

Boiler water invariably contains mineral salts and gases in solution. Dissolved gases attack metal boiler surfaces, causing extensive corrosion and pitting. The dissolved minerals combine chemically under pressure and temperature to form scale deposits. These reactions, if not controlled, can rapidly reduce boiler efficiency.

Deposits and corrosion are controlled by adding chemicals directly to the boiler as shown in Figure 7.3. Chemical feed systems are frequency used to store and inject the chemicals in precisely controlled amounts (see Figure 7.4).

For best results, all chemicals for internal boiler water treatment must be fed continuously and at proper injection points. Shot feeding is not recommended for any steam generating plant, particularly a modern one operating at higher heat transfer rates.

The shot method is used only with closed heating systems, particularly if a barrier-type corrosion inibitor is used. All chemical feedlines should have an open-end discharge rather than a perforated line such as is used for continuous blowdown collection.

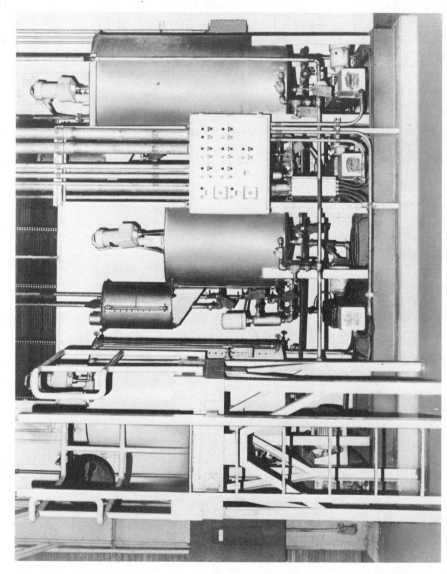

Figure 7.3. Chemical feed system for treating boiler feedwater.

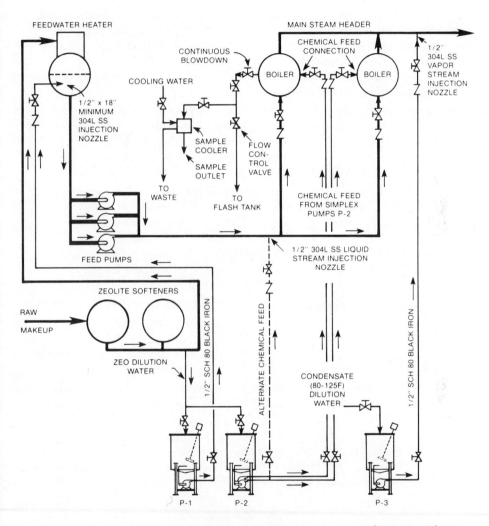

Figure 7.4. Typical piping for boiler system chemical control. [Reprinted from *Betz Handbook of Industrial Water Conditioning*, Betz Laboratories, Inc., Trevose, PA, 7th ed. (1976).]

No treatment containing orthophosphate should be fed through the boiler feedline since it may produce calcium phosphate feedline deposits. Orthophosphate should be fed directly to the boiler drum. Polyphosphates must not be fed through the boiler feedline when economizers, heat exchangers or stage heaters are part of the pre-boiler system. If the pre-boiler system does not include such equipment, polyphosphates may be fed to the feedwater providing total hardness does not exceed two ppm.

When caustic soda is used for maintaining boiler water alkalinity, it is usually fed directly to the boiler drum. When caustic is used to adjust the feedwater pH, it is normally injected into the storage section of the deaerating heater.

Industrial or refined-grade uncatalyzed sodium sulfate may be mixed with other chemicals. Preferred injection is to the storage section of the deaerating heater at a point where mixing with the discharge from the deaerating section will occur.

Sulfites at feeding strength are corrosive to mild steel tanks because of differential oxygen concentration at the liquid level. Mild steel can be used if the pH of the sulfite solution is raised to 11 or above with caustic soda. Approximately one pound of caustic is needed for 50 gallons of 5% solution.

Sulfite shipped as liquid concentrate is usually acidic, and when fed neat (undiluted), it will corrode a stainless steel tank at the liquid level. The tank must be polyester fiberglass or polyethylene. Lines may be PVC or 316 stainless steel.

1) Catalyzed Sulfite

Catalyzed sulfite must be fed alone and continuously. Mixing catalyzed sulfite with any other chemical will deactivate the catalyst. For the same reason, catalyzed sulfite must be mixed with only condensate or demineralized water. To protect the entire preboiler system, including any economizers, catalyzed sulfite should be fed to the storage section of the deaerating heater.

A steel tank is not satisfactory with this material because caustic soda cannot be used to adjust the pH as can be done when regular sulfite is fed. Materials of construction for feed equipment are the same as for regular sulfite.

2) Hydrazine

Hydrazine is compatible with all boiler water treatment chemicals except organics and nitrates. However, it is good engineering practice to always feed hydrazine alone. Like sulfite, feed is usually continuous into the storage section of the deaerating heater. Materials of construction are the same as those specified for sulfite. Human toxicity precautions should be maintained when working with hydrazine.

3) Chelants

All chelant treatments must be fed to the boiler feedline using an injection nozzle at a point beyond the discharge of the boiler feed pumps. If heat exchangers or stage heaters are present in the boiler feedline, the injection point should be at the discharge of these units.

At solution strength and elevated temperatures, chelating agents can corrode mild steel and copper alloys so that 304 stainless steel is recommended for all equipment.

Chelants should never be fed directly into a boiler since localized attack of boiler metal can occur. Also, since stainless steel chemical lines would be required, chloride stress corrosion could cause the chelant feedline to fail inside the boiler.

4) Filming Amines

All filming amines should be fed into the main steam header at a point which will allow proper distribution. Some systems may need more than one feedpoint to accomplish this. In every case, the steam distribution should be carefully investigated and feedpoints established to prevent any stagnant areas similar to the ones that sometimes occur at the dead end of a header.

Also, many amines must be mixed with condensate, or in some cases, demineralized water. Water containing dissolved solids cannot be used because the solids will contaminate the steam and may produce unstable amine emulsions.

Steel tanks have been used to feed filming amines, but some corrosion does occur above the liquid level. The use of stainless steel is recommended. The equipment specifications are the same as for regular sulfite except that a vapor-type injection nozzle is required.

5) Neutralizing Amines

Neutralizing amines are usually applied to the pre-boiler system (storage section of deaerating heater) or directly to the boiler. No special precautions are needed. A standard packaged steel pump and tank can be used for feeding.

6) Phosphate-Dispersants

These types of treatment are employed in a majority of applications. As with the neutralizing amines, standard steel feeding equipment can be used and the general guidelines for all chemical additives may be followed. Injection should be to the boiler drum.

C. External Boiler Feedwater Treatment

Feedwater from the condensate and make-up source is fed into the feedwater heater, where entrained oxygen and carbon dioxide are removed under temperature and pressure. Although this water has already been treated, it still contains impurities in the form of dissolved minerals and slight traces of entrained gases.

Dissolved minerals form scale deposits on metal boiler surfaces, piping and tubing; continued attack by oxygen and other entrained gases causes boiler system components to corrode. Chemical feed systems add precisely measured concentrations of the following chemicals to the boiler's feedwater side (see Figure 7.4) to help minimize corrosion, scaling and embrittlement:

oxygen scavengers-NaS, Na_2SO_3, N_2H_4;
alkalies-NaOH;
nitrates-$NaNO_3$.

D. Demineralizer Regeneration

One or more demineralizers (Figure 7.5) treat cold water before it enters a boiler for steam generation. Each unit uses cationic and/or anionic resins to remove soluble inorganic salts from the water by adsorption or ion exchange. The cation resin exchanges hydrogen for raw water cations; the anion resin exchanges hydroxyl anions for highly ionized anions. With continued operation, these resins become saturated with the accumulated salts and must be periodically regenerated. Regeneration is accomplished by flushing the resins with acids and alkaline solutions. Because of the precision required in this process, metering pumps are used to inject the following regenerating acids and alkalies:

acids – H_2SO_4 (cation);
alkalines – NaOH, NaCl (anion).

During cationic unit regeneration, sulfuric acid is added to the resin bed according to the manufacturer's instructions. As a rule of thumb, a 2% acid solution is metered into the unit at 0.5 to 0.1 gpm/cu ft of resin. Accurate injection is important because high acid concentrations and/or low flow rates allow calcium sulfate to precipitate on the resin bed. Subsequent to initial acid introduction, the concentration can be increased to 4% and then to 6% to improve regeneration efficiency and to reduce sodium leakage during the subsequent service cycle.

The anionic unit is similarly regenerated, except that a regenerant solution containing 0.5 to 1.0% caustic soda (NaOH) and 10% sodium chloride (NaCl) is

Figure 7.5. Typical make-up water demineralizer.

injected at approximately 1.5 to 3 gpm/sq ft of resin. Although a solution containing only sodium chloride may be used, the addition of caustic results in higher operating capacity and more consistent effluent quality.

E. Process Additives

The liquid raw materials used in chemical processing usually go through several process operations before they are converted into the finished chemical product. Depending on the nature of the end product, additional chemicals must often be injected directly into the process stream to control or to stimulate chemical reactions.

Chemical additives must be introduced at precisely controlled rates; thus metering pumps are often used to meet quantity and concentration requirements. Typical process additives include:

   corrosion inhibitors;
   dispersants;
   acids $-$ H_2SO_4.

1) Corrosion Inhibitors

The obvious solution to acid attack is to remove or neutralize the acid components of the corrosive medium. Accordingly, water-soluble alkalies such as sodium hydroxide (NaOH) and sodium carbonate (Na_2CO_3) and particularly ammonia (NH_3) or ammonium hydroxide (NH_4OH), have long been used for corrosion control. Low-molecular-weight amines, such as morpholine, are also used as neutralizers.

A disadvantage of neutralizing chemicals is that their cost is proportional to the amount of acid corrodent neutralized. Furthermore, close control of the neutralizer is necessary or pH may be raised high enough to cause formation of calcium and magnesium salt scales. Danger to copper and its alloys exists when pH values over 7.5 are reached by using ammonia and many of the neutralizing amines that form soluble complexes with copper.

This danger has been overcome by use of oil-soluble amines of intermediate molecular weight. These materials increase pH at a lower rate as the acids are neutralized. An additional advantage is that hard, crystalline deposits such as ammonium chloride (NH_4Cl) are not formed. NH_4Cl can form by the neutralizing reaction in the vapor phase of NH_3 and HCl. It can contribute to a fouling problem, since no water is present to dissolve the salt. Addition of water for this purpose presents an additional operational requirement.

Film-forming inhibitors are also used in refineries and chemical plants. These are surface-active materials also known as semipolar or reverse-wetting inhibitors. The inhibitor consists of a polar head, such as a fatty acid or amine group, which is attracted to the metal, and a long hydrocarbon tail, which is repelled by the metal. The nature of the polar group is important in determining adsorption on the metal. The length and configuration of the tail determine relative oil and water solubility and other properties. An important function of the hydrocarbon tail is to attract some of the oil or hydrocarbon constituents flowing past the metal. This oil, enmeshed in the tail as shown in Figure 7.6, provides a mechanical barrier to attack of the aqueous corrodents on the base metal.

A major advantage of film-forming over neutralizing inhibitors is that a stoichiometric relationship between amount of film-former and amount of active corrodent is not required as with neutralizers. Requirements for film-forming inhibitors depend on the total amount of fluids contacting the surface to be protected, and minimum fluid concentrations must be maintained to effect a tight, protective film on the metal.

The combination of adsorbed inhibitor with the oil it attracts to it acts as a physical barrier similar to a protective coating. But this barrier is self-healing because of continuous addition of the inhibitor. This is an advantage over paints, coatings and linings where breaks in the barrier require shutdown for maintenance and repair.

A combination of neutralizers and film-forming inhibitors is usually more effective and economical than is either one alone. Ammonia or other neutralizing amine is fed continuously with a filming inhibitor. The conditions that determine the effectiveness of these systems are temperature, compositions, relative amounts of the aqueous and oil phases, and inhibitor concentrations.

The treatment's ultimate value is determined by field application and monitoring. In addition to the factors mentioned above, density, volatility and solubility of a proposed inhibitor must be considered. These properties can determine the point of addition to the system, and carrier type.

One of the most important uses of inhibitors of the neutralizing and/or filming type is in the overhead systems of crude stills in which the lower pH of condensed water would normally lead to high corrosion rates of steel equipment. A typical system is shown in Figure 7.7.

2) Hydrogen Blistering and Embrittlement

In addition to the overall attack as measured by uniform weight loss or other methods, a more subtle form of damage is experienced when sulfur compounds are present. This damage is caused by hydrogen blistering or embrittlement resulting from migration of atomic hydrogen through the metal.

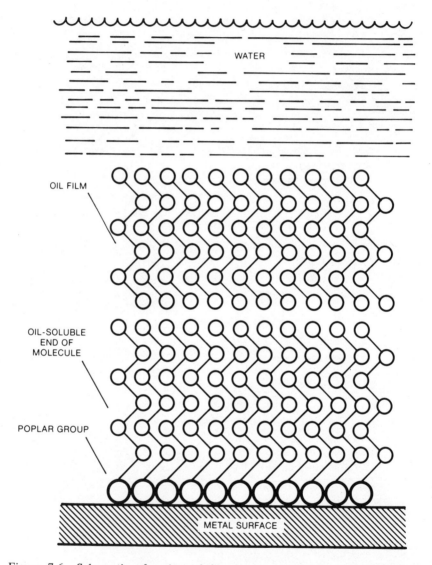

WATER

OIL FILM

OIL-SOLUBLE
END OF
MOLECULE

POPLAR GROUP

METAL SURFACE

Figure 7.6. Schematic of action of film-forming inhibitor. [Reprinted from *Betz Handbook of Industrial Water Conditioning*, Betz Laboratories, Inc., Trevose, PA 7th ed. (1976).]

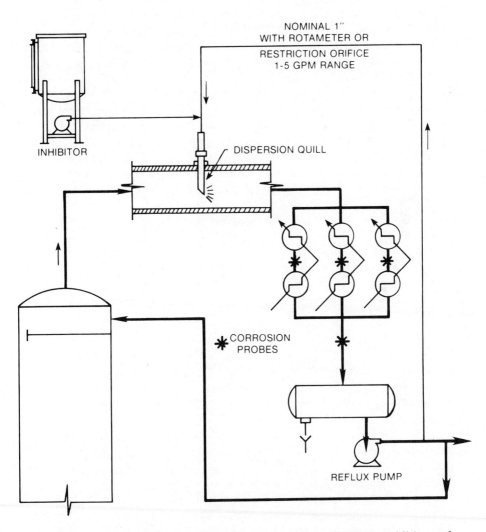

Figure 7.7. Schematic of overhead system with points of inhibitor addition and probe locations. [Reprinted from *Betz Handbook of Industrial Water Conditioning*, Betz Laboratories, Inc., Trevose, PA 7th ed. (1976).]

Among the various methods of reducing corrosion, both overall and blistering/embrittlement in the presence of cyanides, is the use of oxygen and polysulfides which convert the detrimental cyanides to less objectionable cyanates or thiocyanates. Another technique is water-washing, which dilutes the concentration of the corrodents and accordingly reduces corrosion rates. These methods require close control of the injection of oxygen or polysulfide, the handling of the corrosive polysulfide solutions and the disposal of polluting wastewater streams.

An attractive alternative to the above methods of combating cyanides is to use film-forming inhibitors developed especially for use in these high pH fluids. In common with other inhibitors, they form an oil-attractive, water-repellent film on the metal surface. This film prevents contact by the corrosive fluids on the base metal, reducing the overall corrosion rate and concomitantly decreasing the total amount of hydrogen (atomic and molecular) formed by the primary corrosion reactions. Accordingly, less hydrogen migrates into the metal and less damage is done. These materials are used effectively in refinery streams, coke oven gas recovery plants and similar applications.

Filming inhibitors are effective because of their high surface activity which causes adsorption at the interface between the metal and liquids. A secondary and undesirable effect is adsorption at either liquid-liquid or liquid-gas interfaces, which can contribute to emulsion or foaming problems. These problems can be minimized by formulation of treating chemicals for specific applications.

3) Refinery/Petrochemical Process Fouling

Process equipment fouling caused by the formation of adherent deposits which interfere with equipment operation is a major problem of the petroleum industry. In many instances, fouling and resulting economic losses can be avoided or greatly reduced by use of special chemicals.

Fouling deposits can be classified as inorganic and/or organic. Inorganic deposits may be very simple in identity and the solution to the problem may be direct.

Inorganic deposits, such as oxides, sulfides and other salts of iron, copper and vanadium, may be present in the original hydrocarbon feed as ash. They may also result from corrosion where fouling is evidenced in units upstream of the fouling problem area.

In many cases, fouling and corrosion may not be independent problems. If a fouling deposit has a large inorganic content, a corrosion problem may exist in the system and may be solved by using inhibitors.

Organic foulants have rarely been completely identified chemically. They are usually high-molecular-weight materials of uncertain composition containing carbon, hydrogen, oxygen, nitrogen and sulfur.

Fouling deposits often result from chemical reactions occurring in process equipment. In petrochemical plants, severe fouling can occur when monomers used as feedstock convert to polymers.

Fouling is evidenced by an increase in pressure drop through pipes, by a reduction of heat transfer coefficients in exchangers, by increased fuel or refrigeration demands, or by complete breakdown or malfunction of units. The economic losses may be severe.

Fouling problems can be solved either by preventing the formation of foulants or, if this is not possible, by minimizing their effects by injecting anti-oxidants, chelants, dispersants, and multipurpose additives.

F. Process Liquids

In chemical and petrochemical processes, raw materials are often combined under high temperatures and pressures to initiate chemical reactions. The liquid products of processing frequently contain significant amounts of suspended or dissolved impurities, and these must be removed by further processing or refining operations.

Each of these operations involves physically moving the process liquid in a continuous stream from one reactor or vessel to another. Because the process liquid stream must flow at a precise rate throughout the system, metering pumps frequently handle such liquids as:

acids — HCl, HNO_3;
ammonia — $NH_3(OH)$;
alcohols.

G. Fuel Additives

A fuel additive is a substance—in either slurry or liquid form—which is added directly to the fuel used to fire a boiler for steam generation.

The additive acts as a catalyst to promote more efficient fuel combustion, minimize corrosion and deposits and improve boiler efficiency by aiding heat transfer.

Because relatively small quantities are required, low-capacity metering pumps are used to inject the following fuel additives:

MgO;
Al_2O_3;
MnO_2.

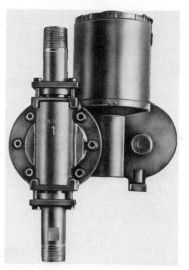

Figure 7.8. Typical fuel oil additive pump featuring tubular diaphragm for improved flow characteristics.

Slurried fuel additives generally consist of a metal oxide in a light hydrocarbon oil and are often less expensive than their soluble counterparts. To handle such additives, a straight-through tubular diaphragm type pump (see Figure 7.8) is frequently specified as part of the total additive feed package. The system should be designed so as to allow periodic flushing with mineral oil.

Soluble additives can be handled by most types of metering pumps. These additives may be added directly to the feed system storage tank when it is filled, or they may be proportioned into the storage tank feed line to eliminate the need for manual addition.

H. System Flushing

The reactions that take place in chemical and petrochemical processing equipment often produce fouling deposits, sludge and clogging. Any combination of these conditions over a period of time will increase pressure drops, reduce heat transfer, increase fuel consumption and, in extreme cases, cause a complete stoppage of processing operations. To properly control fouling and deposits, the system components must be periodically cleaned and flushed. This procedure breaks up and removes the undesirable deposits, restoring the system to its original efficiency.

Another application of system flushing is in batch processing of two or more different products. The system must be flushed to remove all traces of the first product and to prevent contamination before the next product is processed. Liquids used in system flushing may be organic or inorganic, acid, alkaline or pH-neutral (depending on the chemical composition of the deposits or residue to be removed). Metering pumps can be used to inject additives into a flushing or cleansing liquid. Typical additives include:

acids — H_2SO_4;
alkalies — NaOH;
dispersants;
corrosion inhibitors.

I. Process Water Treatment

Water is used extensively in many process applications by chemical and petrochemical producers. A typical chemical processor may use water to convey, grade, wash, crush, dissolve, or absorb a raw material. Water is frequently a basic component of the finished chemical product.

After being subjected to heat, pressurization and blending, chemical ingredients may undergo additional processing to extract unwanted solid, liquid or gaseous impurities. These impurities are then processed further to recover desired ingredients.

The process water resulting from these operations contains high concentrations of dissolved chemicals, suspended solids and entrained gases. Since much of this process water is recycled for use in other stages of processing, it must be treated to restore it to as pure a state as possible.

Process water treatment involves addition of specific chemicals in precise amounts to minimize the harmful effects of corrosion, hydrogen blistering and embrittlement and fouling before the water can be recirculated or discharged. Metering pumps often deliver chemicals that change the aggressive nature of this environment. Chemical treatment offers several advantages when compared with the metallurgical approach or the use of anti-corrosion barriers:

Initial expenditures are generally reduced.
The cost of chemicals is a maintenance or operating expense rather than an operating expenditure.
Greater flexibility is possible, since the types and quantities of chemicals used can be varied according to the composition and requirements of each system.

Due to the large number of variables, treatment methods and systems are designed individually to provide optimum results at minimum cost.

II. ELECTRIC POWER GENERATION

Major metering pump applications in the electric power generation industry include:

cooling tower water treatment (see Section I-A);
internal boiler water treatment (see Section I-B);
external boiler feedwater treatment (see Section I-C);
demineralizer regeneration;
fuel additives;
wastewater treatment.

A. Demineralizer Regeneration

Figure 7.9 shows a disc diaphragm metering pump used by a major utility to regenerate cationic resin beds. Figure 7.10 shows a typical water quality analysis lab, which provides signal input to automated chemical feed systems. See Section I-D for a discussion of demineralizer regeneration.

B. Fuel Additives

See Section I-G.

MgO;
MnO_2;
Al_2O_3.

III. PETROLEUM REFINING AND NATURAL GAS PROCESSING

Major metering pump applications in the petroleum refining and natural gas processing industries include:

cooling tower water treatment (see Section I-A);
internal boiler water treatment;
process additives;
process liquids;
system flushing;
process water treatment;
wastewater treatment.

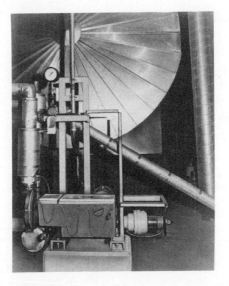

Figure 7.9. Disc diaphragm metering pump is used to inject sulfuric acid during cationic resin bed regeneration.

Figure 7.10. Part of a typical water quality analysis lab, which provides feedback to a chemical feed system to permit automatic maintenance of water quality.

A. Internal Boiler Water Treatment

Refinery processes require tremendous amounts of heat—approximately 675,000 BTU per barrel of crude oil. Pressurized steam supplies the heat needed to separate crude oil into its many hydrocarbon components and to refine these components into finished products such as gasoline, fuel oil, asphalts and kerosene. See Section I-B for treatment details.

B. Process Additives

Process equipment corrosion and fouling are major problems in petroleum refining and natural gas processing. These problems are initially encountered with "sour gas" at the well head; they are subsequently encountered in other stages of processing operations, from crude oil fractionation through polymerization and blending.

The hydrocarbon process stream contains dissolved gases, acids and salts. These substances, under high temperatures and pressures, tend to be extremely corrosive on metal surfaces. They also interact to form organic deposits which foul the process system.

To minimize corrosion and fouling, chemical additives are introduced directly into the process stream. These additives may be metal deactivators, antioxidants, dispersants or inhibitors. All are intended to reduce fouling and corrosion.

The amount of each additive introduced is critical, and they must be delivered at carefully controlled rates which vary with the process and hydrocarbon involved. Metering pumps satisfy these requirements, controlling the amounts and flow rates of the following:

> Corrosion inhibitors;
> Dispersants, antifoams;
> Aluminum chloride — $AlCl_3$.

C. Process Liquids

In the many stages of petroleum refining operations, crude petroleum is separated and converted into a number of hydrocarbon products. Among these are gasoline, kerosene, fuel oil, lubricating oil, waxes, asphalts and liquified petroleum gases.

Each of these hydrocarbons is a process liquid, and each must be physically moved through further processing and refining operations after being separated. In many of these refining processes, metering pumps are used to inject control agents into the process liquid stream:

Figure 7.11. Metering pumps are an integral part of many process octane analysis systems.

Acids — H_2SO_4, HFl;
Naphthas;
Paraffins.

Metering pumps are also used in octane analysis systems, such as the one shown in Figure 7.11. In this application, the pump adds a precise quantity of the final product for exothermic reaction to permit determination of octane.

D. System Flushing

The feedstock, intermediate hydrocarbons and refined products which flow through petroleum refining and natural gas process systems contain varying amounts of organic and inorganic compounds. These compounds may be suspended, dissolved or entrained, and under the extreme temperatures and pressures required in refinery operations tend to foul the processing system with sludge or other deposits.

Such undesirable deposits in pipes, tubes and vessels will decrease the flow of process liquids, increase fuel consumption and lower processing efficiency in general. See Section I-H for a discussion of system flushing methods.

E. Process Water Treatment

Water is an important factor in many petroleum refinery process operations. Refining involves two basic operations:

1. Physical separation processes, such as distillation and extraction, which separate crude petroleum into many hydrocarbon components having different boiling points and solubilities.
2. Chemical conversion processes, such as catalytic reforming and hydrotreating, which change the molecular structure of the hydrocarbon components.

Many of these operations are interrelated, and process water is recycled and recirculated between them. Large quantities of process water result when water is separated from hydrocarbons during fabrication, distillation and catalytic cracking. Steam condensate from the fractionater is another source of process water.

The resulting process water may contain significant amounts of sulfides, chlorides and organic compounds which tend to form troublesome emulsions, causing corrosion on metal surfaces. To minimize these conditions, process water must be chemically treated before it is recycled.

Metering pumps add the following required chemicals to the process water:

emulsion breakers;
corrosion inhibitors.

IV. WOODPULP AND PAPER PROCESSING

Major uses of metering pumps include:

cooling tower water treatment (see Section I-A);
internal boiler water treatment;
external boiler feedwater treatment (see Section I-C);
demineralizer regeneration (see Section I-D);
process additives;
process liquids;
process water treatment;
wastewater treatment.

A. Internal Boiler Water Treatment

Pressurized steam is used in many stages of pulp and paper processing operations, including pulping, digesting, refining, heating liquor and drying. See Section I-B for a discussion of treatment methods.

B. Process Additives

The raw materials for woodpulp processing are wood and water. But in the process of converting wood into pulp into paper products, chemical additives are necessary to soften woodchips, loosen pulp fibers, correct pH levels or process liquors, aid precipitation, thicken or dilute pulp and facilitate the bleaching process. They are also used to control corrosion, scale deposits and bacterial growth.

Additives are introduced at many stages of processing, under varying temperatures and pressures. Timing may be critical, and the amount and concentration of each additive must be carefully controlled. Metering pumps are used to deliver the following process additives:

dispersants;
anti-foams;
biocides;
retention aids;
acids;
alkalies.

C. Process Liquids

The manufacture of paper products from wood involves grinding wood chips into pulp, mixing pulp with water, using heat, pressure and chemical reactions to separate pulp fibers, screening and filtering the pulp and, finally, removing water by draining, pressing and drying. A typical paper machine is shown in Figure 7.13.

Each step of the operation brings about physical or chemical changes in the pulp, and process liquids are involved in virtually every step. As pulping, digesting and refining operations progress, the pulp's viscosity changes. During the pulping process, water content may range from 99% to 6%.

The process liquids must be conveyed from one processing operation to

Figure 7.12.   Logs stored in the woodyard await conversion into woodpulp and, eventually, paper.

Figure 7.13.   Typical paper machine combines draining, pressing, and drying functions to convert woodpulp to paper or board.

another either in batches or in a continuous stream. Because of the flow of these liquids must be rigorously controlled, metering pumps are used to inject the following additives:

white liquor $-$ Na_2 $+$ NaOH;
salt cake $-$ Na_2SO_4;
green liquor $-$ Na_2SO_3.

D. Process Water Treatment

Water is used in virtually every stage of pulp and paper processing. It helps dissolve chemicals in pulping and bleaching processes, transports pulp fibers, washes wood pulp, removes excess chemicals, and is a component of the finished product. Process water in the form of black, green, brown or white liquor is essential in all process operations. The process water used in pulp and papermaking operations contains high concentrations of organic and inorganic solids, as well as entrained gases. Since the water is reused in many stages of processing it must be treated to minimize the harmful effects of pollutants.

The conditions required for paper manufacturing are quite conducive to bacterial growth. Fiber, cellulose and other organic matter extracted from wood pulp together with warm, oxygenated water provide an excellent medium in which bacteria rapidly multiply, forming slime deposits and corroding metal surfaces. Slime and scale deposits also form as the process water is evaporated and circulated.

1) Pulping Process

Influent treatment methods in which metering pumps are employed include chlorination, coagulation, flocculation, sedimentation, filtration and corrosion and scaling protection.

In the woodyard, thermal and biological factors contribute to storage losses. If water sprays are specified to reduce thermal losses, biocides are generally proportioned into the water to curb biological attack.

Bleached pulp mills often inject polyelectrolytes into influent surface water to enhance coagulation and flocculation, which, in turn helps to control turbidity and color. To reduce the calcium hardness of well water, a softening process that includes injection of acidic solutions is often necessary. Scaling is subsequently controlled through use of polyacrylates.

Metering pumps also contribute to kraft pulping processes. Chemical delignification of the fiber bundles entails use of a concentrated sodium hydroxide/sodium sulfide solution. Metering pumps are used to combine these two chemicals in the proper proportion and to inject the necessary amount of the resultant solution into the digesters as required. Prior to causticizing, which takes place later in the process, addition of anionic polyelectrolytes to the green liquor dramatically increases the settling rate of such undesirable impurities as iron, carbon and silica compounds. This facilitates green liquor clarification. Removal of calcium carbonate particles from the white liquor is expedited in a similar manner. Polyelectrolytes are also frequently metered into the mud washers to improve settling rates and flow characteristics.

In the sulfite pulping process, magnesium hydroxide is fed into the spent liquor to aid in the recovery of reusable chemicals. Dispersants are also added to the liquor to help prevent scaling. Pitch deposition, which commonly accompanies mechanical pulping, is controlled through pH adjustment and charge neutralization. To aid in pitch control, alum and caustic may be used during grinding or disc refining; however, various dispersants are frequently substituted for alum during these processes to avoid the harmful effects of excessive aluminum sulfate.

Metering pumps also play an important role in the semichemical pulping process. Here, a solution consisting of sodium hypochlorite and hydrogen peroxide oxidizing agents is injected to increase brightness. During the alternative peroxide bleaching process, a 35% or 50% by weight solution of hydrogen peroxide is substituted.

Reducing agents are sometimes used in place of or in conjunction with oxidizing agents. Among these are zinc and sodium hydrosulfite solutions, which are generally added to the pulp after grinding.

Multistage bleaching processes, used for pulps of high brightness, entail precise addition of solutions of chlorine, chlorine dioxide, sodium hypochlorite, sulfurous acid, caustic, and/or sulfur dioxide. Metering pumps are used to add some or all of these solutions in various sequences and quantities.

2) Papermaking

The papermaking process uses approximately 40,000 gallons of water to produce one ton of paper. For obvious reasons, it is desirable to recycle as much of this water as possible. This dictates careful addition of numerous chemicals which frequently involves use of metering pumps and chemical feed systems.

Various biocides are injected to aid in control of fouling caused by fungi, bacteria, and biological slime. Nonmicrobiological deposits—pitch and scale—can also be chemically controlled. Alum and polyphosphate solutions are two of the pitch control agents that are commonly metered into the process water. Dispersants are preferred for scale control; these include synthetic organic surfactants, lignosulfonates, tannins, polyelectrolytes and polyphosphates.

To aid in retention, metering pumps add certain flocculants to decrease the amount of fiber and filler that are lost during sheet formation. These include high molecular weight cationic polymers and synthetic polymeric aids and alum solutions.

Process water must also be treated to help prevent corrosion; this is the domain of the corrosion engineer. In general, the acidic pH range of the water increases the potential for corrosive damage; therefore, alkalies are frequently metered into the process water. Numerous other solutions may also be metered or added to reduce corrosive wear.

V. FOOD AND BEVERAGE PROCESSING

Major metering pump applications in the food and beverage processing industries include:

cooling tower water treatment (see Section I-A);
internal boiler water treatment (see Section I-B);
external boiler feedwater treatment (see Section I-C);
demineralizer regeneration (see Section I-D);
process additives;
process liquids;
system flushing;
process water treatment;
wastewater treatment.

A. Process Additives

Process additives are used more frequently in food and beverage processing than in any other major industry. They are added to process water and liquids to initiate chemical reactions, remove undesirable impurities, separate the various components of raw food material, reduce foaming, retard bacterial growth and food spoilage, and enhance the flavor, color or odor of the product.

Some additives are used only in intermediate stages of processing, while others become ingredients of the final food product. All additives must be carefully proportioned; too much will unfavorably affect taste, color or odor; too little will fail to produce the desired effect. Metering pumps, therefore, are used to inject the following additives:

biocides;
antifoam agents;
acids — H_2SO_4;
alkalies — NaOH.

B. Process Liquids

The final products of food and beverage processing operations may be solid, semi-solid or liquid in physical form. In virtually every segment of the food processing industry, however, liquids are either the final product, ingredients of the final products or are essential to intermediate processing steps.

Depending on the type of final product, process liquids may range in viscosity from water to molasses, in pH value from acidic to alkaline, and in process temperature from 0° to 100°C.

Process liquids are heated, blended, refined, distilled and filtered. Figure 7.14, for example shows the process by which beer is brewed; Figure 7.15 shows the control room from which the brewing process is monitored. Through all these processing operations, liquids must be physically moved from one phase to another, in a continuous stream.

The metering pumps shown in Figure 7.16 are used to inject a diatomaceous earth slurry used in microfiltration processes. This filtering procedure is widely used by many breweries and wineries as well as for polishing water already clarified by other means. A packed plunger type metering pump is used to inject the abrasive slurry. Prior to the start of the filtering process, the pump coats the septum (or filter base) with a predetermined amount of diatomaceous earth. Additional diatomite slurry is added intermittently during the process until the accumulation of matter generates excessive back pressure. Then the filter must be shut down to permit the bed to be backwashed by a centrifugal pump. In order to minimize the need for backwashing and yet optimize filtration, the slurry must be precisely proportioned against high system pressure.

C. System Flushing

Food and beverage processing systems are subject to corrosion of metal surfaces and formation of scale deposits resulting from food spillage and leakage. In addition, biological activity is a unique problem in this industry. The nutrients in food products tend to promote rapid microbacterial growth in processing equipment.

To minimize scale deposits and bacterial growth in food processing equipment—and especially to prevent contamination and safeguard public health—all components of food processing systems must be frequently flushed out with cleansing and purifying liquids. These are frequently mixed by metering pump systems, which proportion the liquids into the flushing main to provide the proper dilution at variable flow rates. Cleansing and purifying liquids include acids and chlorine solutions.

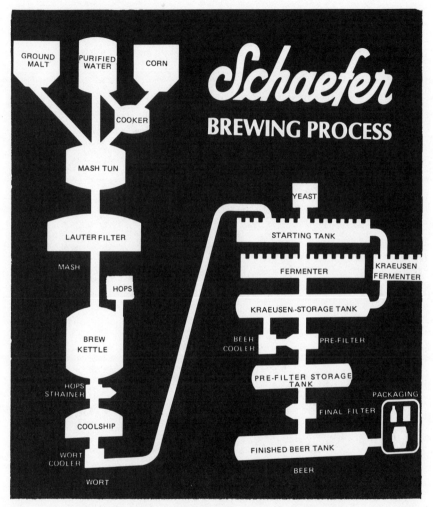

Figure 7.14. Typical brewing process. Metering pumps are used to add a diatomaceous earth filter aid slurry during pre- and final filtration stages.

D. Process Water Treatment

Process water is indispensible in food and beverage processing operations. The major processing applications of water are to:

Wash raw food materials to get rid of such solid wastes as soil, plant fiber and unwanted organic matter.

Figure 7.15.   The brewing process shown is controlled by the master brewer.

Figure 7.16.   Abrasive diatomaceous earth is handled by packed-plunger metering pumps. Metering accuracy is essential in this application, as too much slurry results in filter blockage; too little results in inadequate purification.

Leach out solid impurities before food products are canned or bottled.

Wash processing equipment.

Cook and process raw food products.

Serve as a vehicle for chemical additives that remove impurities causing undesirable taste, color and odor effects.

Sanitize and sterilize containers before filling.

Heat and cool cans and jars after sealing.

Physically move products via flume from one process to another.

Process water from these operations is usually laden with chemicals, such as chlorine, detergents, biocides, lime, foam dispersants and pH control agents. Because process water is usually an excellent medium for bacterial growth, and the direct connection between food products and public health, it is closely monitored by local, state and federal governmental agencies.

The U.S. Department of Agriculture (USDA) and the Food and Drug Administration (FDA) rigorously regulate processing operations. To comply with government regulatory codes, food and beverage processors must scrupulously control process water composition. Metering pumps are used to add chemicals which restore process water to government-mandated levels of chemical, physical and thermal composition. Figure 7.17 shows a typical system used for proportioning sanitizing chemicals into cleanup water.

1) Corrosion Control

Polyphosphates and soluble silicates are widely used in the industry as corrosion inhibitors. Sodium silicates have been applied as corrosion inhibitors for many years. When sodium silicates are combined with polyphosphates, a highly beneficial synergistic effect is produced: corrosion is inhibited more effectively than with either inhibitor alone. Metering pumps play an integral role in this process.

Another inhibitor system combines polyphosphate and zinc. One disadvantage with this type of treatment is that control is critical. At alkaline pH, zinc orthophosphate becomes insoluble. If control is poor and inhibitor is overfed or pH is not closely controlled, zinc orthophosphate precipitation problems arise, can spotting occurs and corrosion protection declines.

The increased chemical deposition poses another problem: when the zinc phosphate deposits in low flow areas such as the cooling basin, it acts as an excellent nesting ground for microorganisms.

Decharacterized sodium sulfite has been used as an oxygen scavenger in the water columns and sterilizer section of a hydrostatic cooker. Since oxygen is continuously being introduced into the system, relatively high (100 parts per million) sulfite concentrations are necessary to drive the oxygen-scavenging reac-

Figure 7.17. Pneumatic metering system proportions liquid sanitizing agent into incoming clean-up water on a stroke basis. This permits compliance with FDA regulations. Storage tank is visible at left.

tion to completion. Eliminating oxygen from the water effectively stifles corrosion.

The sterilizing section of a cooker must be protected from attack by carbonic acid and possibly oxygen. Filming amines, neutralizing amines/hydrazine and ammonia/hydrazine are among the chemicals injected for such protective treatment. The use of filming amines is the most widely accepted way to treat steam in cookers because results are consistently excellent. Filming amines form an impervious, non-wettable film on metal surfaces which acts as a protective barrier against oxygen and carbon dioxide attack. Hydrazine is often fed in combination with neutralizing amines because it scavenges oxygen. However, neutralizing amines must be fed in a stoichiometric ratio to remove carbonic acid. For example, about 3.0 parts per million of cyclohexylamine (40 percent) are necessary for each part per million of carbon dioxide, to elevate the pH of the condensate to 7.0. Consequently, depending on the level of carbon dioxide, the cost of this program can be substantial.

VI. MISCELLANEOUS AND COMMERCIAL APPLICATIONS

The applications previously covered represent the broad spectrum of metering pump end uses. Other industries in which the pumps are used include mining

TABLE 7.3

Typical Metering Pump Chemicals by Industry

Potable/Waste Water Treatment
Calcium carbonate (lime slurry)
Calcium hydroxide (slaked lime)
Chelants
Ferric chloride
Ferrous sulfate
Lime slurries
Potassium aluminum sulfate
Sodium fluoride
Sodium hydroxide (caustic soda)
Sodium hypochlorite
Sulfuric acid

Fertilizer and Agricultural
Ammonium nitrate
Ammonium phosphate
Ammonium sulfate
Hydrochloric acid
Potassium carbonate (potash)
Sodium hydroxide
Trisodium phosphate

Textile
Acetone
Chlorine
Dyes
Lubricants
Modified sodas
Sodium alginate

Petrochemical
Acetaldehyde
Acetic acid
Aluminum chloride
Benzaldehyde
Bromine
Cresylic acid
Ethyl acetate
Ethyl mercaptan
Hydrofluoric acid
Iodine solutions
Naphthalene solutions
Oleum

Power Generation
Algicides (various)
Amines (various)
Ammonium hydroxide (aqueous
  ammonia)
Fuel oil additives
Morpholine
Sodium hydroxide (caustic soda)
Sodium phosphate
Sodium sulfite

Plastics
Acetone
Acrylonitrite
Allyl chloride
Copper acetate
Hardeners/catalysts
Resins

Food and Beverages
Acetic acids
Anti-foamants
Beet sugar liquors
Benzoic acid
Calcium hydroxide
Diatomaceous earth
Dyes
Ferrous sulfate
Gum arabic
Phosphoric acid
Sodium hypochlorite
Sorbic acid

Plating
Acetaldol
Barium cyanide
Chromic acid
Copper sulfate
Sodium hydroxide
Sodium hypochlorite

Source: Milton Roy Company

Figure 7.18. Mechanically actuated diaphragm metering pumps are integral parts of many commercial swimming pool chlorine feed systems.

and mineral processing, textile manufacturing, fertilizer production, agriculture and horticulture and plating. A summary of the chemicals that are metered within these industries is included in Table 7.3. Because many of these applications are highly specialized, it is not possible to detail them here.

Metering pumps are also used in a wide variety of commercial applications, ranging from "water doctor" supplied commercial cooling tower treatment systems to automatic chlorination systems for swimming pools. Because commercial applications are typically intermittent, mechanically actuated diaphragm type metering pumps offer a cost-effective method for precise injection of many types of additives; they also offer adequate reliability for these noncontinuous applications.

Figure 7.18 shows a mechanically actuated diaphragm pump used to inject sodium hypochlorite into swimming pool water. Metering pumps are also used in laboratories, car wash equipment and low-pressure commercial boilers.

8

WASTEWATER AND POTABLE WATER TREATMENT

In order to adequately assess the role of metering pumps in wastewater and potable water treatment, this chapter begins with a discussion of general treatment practices. Specific industrial practices are then outlined; finally, the use of metering pumps in treating potable water is explained.

I. OVERVIEW

Wastewater is the effluent discharged from municipal, commercial and industrial sources. In all municipalities, the wastewater which must be treated contains sewage produced by community and household activities (and in some cases storm water). In larger, more industrialized cities, sewage composition may be affected by wastewater discharge from various industrial sources. Chemical processing, food processing and metal treatment plants, for example, each add different pollutants to municipal wastewater.

A typical municipal wastewater treatment facility treats effluent containing organic and inorganic pollutants. These may be suspended or dissolved, and may vary in viscosity, corrosiveness, abrasiveness and toxicity. The government closely regulates the quality of wastewater discharged into the environment by municipal wastewater treatment plants. Wastewater must be physically and chemically treated to reduce all suspended and dissolved pollutants to environmentally safe, acceptable levels. Treatment proceeds in sequential stages, usually involving:

1. Primary treatment—Raw wastewater passes through screens and filters to remove larger solids and heavy grit particles. It is aerated to reduced the concentration of anaerobic bacteria, and clarified to remove suspended solids.
2. Secondary treatment—Wastewater, with most suspended solids removed, is pumped through trickling filter or activated sludge beds. This step removes most biodegradable organic matter. Various

biological reactors may be added to precipitate or coagulate undesirable elements. The wastewater is again clarified to remove suspended solids.

3. Tertiary treatment—Wastewater is filtered and chemically disinfected to neutralize any remaining harmful organisms or pollutants. It is once again clarified and excess chlorine is removed. The purified, treated water is then discharged into the environment.

The most commonly used water treatment chemicals are listed in Table 8.1. For each, the recommended concentration and material of construction for wetted parts is shown.

A. Clarification of Industrial Wastewater

Clarification entails coagulation, flocculation, and sedimentation. Coagulation brings particles together through charge neutralization; once neutralized, these particles are further agglomerated (flocculated) and physically removed by sedimentation. Although physical sedimentation is used for removal of larger suspended solids, coagulation and flocculation permit removal of smaller suspended material.

In-line clarification involves filtration of water for removal of coarse particles, followed by the precise addition of the coagulant chemicals or polymers shown in Table 8.2. Any or all of these chemicals are also used in conventional clarification equipment and upflow clarifiers. The coagulation-sedimentation process requires three distinct physical-chemical operations:

1. Rapid mix for coagulation.
2. Moderate mix for flocculation.
3. Floc and water separation.

In general, successful coagulation depends upon:

1. The presence of the minimum quantity of chemicals necessary to form an insoluble floc. (Under-treatment results in carry-through of solids. Over-treatment wastes chemicals and results in carry-through of reagent.)
2. The presence of strong acid ions, such as chloride and sulfate. (These ions neutralize positive charges on the suspended particles and cause coalescence.)
3. The control of reaction pH within a specific range. (For each coagulant and water, there is an optimum pH zone for maximum coagulation.)

TABLE 8.1

Commonly Metered Treatment Chemicals

Chemical Name	Symbol	Use	State	Concentration Normally	Typical Material of Construction
Alum	$Al(SO_4)_3$	Coagulation, settling	Slurry	10%	316 SS
Hydrated lime	$Ca(OH)_2$	Coagulation, settling	Slurry	10%	316 SS
Soda ash	$NaCO_3$	Softening agent	Slurry	5%	CI/Steel/SS
Polyelectrolytes	—	Settling	Viscous liquid	2%	CI/Steel/SS
Ferric chloride	$FeCl_3$	Coagulation, settling	Liquid	45%	Plastic
Ammonia	NH_4OH	Nitrogen source	Liquid	100%	CI/Steel/SS
Phosphoric acid	H_3PO_4	Phosphorous source	Liquid	Conc.	316 SS
Sodium hydroxide	$NaOH$	Precipitator	Liquid	20–50%	316 SS
Sodium hypochlorite	$NaOCL$	Disinfection	Liquid	(12% Cl_2)	Plastic

A large variety of chemicals, such as biological additives, clarifying agents and toxicity neutralizers, are precisely metered into wastewater using metering pumps. Some of the more commonly employed chemicals are shown here.

TABLE 8.2

Common Coagulants and Coagulant Aids Handled by Metering Pumps

Name	Formula	Commercial Strength	Grades Available	Weight lb/cu ft	Suitable Handling Materials	Miscellaneous
Aluminum Sulfate	$Al_2(SO_4)_3$ $18H_2O$	17% Al_2O_3	Lump Powder Granules	Powder, 38–45 Other, 57–67	Lead Rubber Silicon Iron	pH (1% solution) = 3.4
Sodium Aluminate	$Na_2Al_2O_4$	55% Al_2O_3	Crystals	60	Iron Steel Rubber Plastics	Stabilized with approx. 6% excess NaOH
Ammonium Alum	$Al_2(SO_4)_3$ $(NH_4)_2SO_4$ $24H_2O$	11% Al_2O_3	Lump Powder	60–68	Lead Rubber Silicon Iron Stoneware	pH (1% solution) = 3.5

Name	Formula	Grade	Form	%	Materials of Construction	Remarks
Copperas	$FeSO_4 \cdot 7H_2O$	55% $FeSO_4$	Crystals Granules	63–68	Lead Tin Wood	Efflorescent
Ferric Sulfate	$Fe_2(SO_4)_3$	90% $Fe_2(SO_4)_3$	Powder Granules	60–70	Lead Rubber Stainless Steel Plastics	Hygroscopic
Ferric Chloride	$FeCl_3 \cdot 6H_2O$	60% $FeCl_3$	Crystals	45–55	Rubber Glass Stoneware	Hygroscopic
Magnesium Oxide	MgO	95% MgO	Powder	25–35	Iron Steel	Essentially insoluble, fed in slurry form
Bentonite		...	Powder	60	Iron Steel	Essentially insoluble, fed in slurry form
Sodium Silicate	$Na_2O \ 3.22SiO_2 \ 41°$ Be		Solution	87	Iron Steel Rubber	Solid grades are available with varied $Na_2O:SiO_2$ ratios

Reprinted from *Betz Handbook of Industrial Water Conditioning*, Betz Laboratories, Inc., Trevose, PA, 7th ed. (1976).

Under normal treatment conditions, most coagulation systems are well buffered in their operating range. A standard proportional feed system controls both the concentration of floc forming reagent and the pH. A constant-rate duplex pump is often employed to effect close control, and in other instances, a proportional feed system meters the floc forming reagent while an automatic control system maintains pH at the desired control point.

Clarification involves introduction of cations (aluminum or iron) that are attracted to negative particle surfaces. A water-soluble organic that ionizes positively may also be introduced (normally a polymer). Inorganic cations may also be injected to form insoluble precipitates that entrap additional particles.

When neutralized particles begin colliding and growing in size, flocculation begins. The speed of this reaction may be increased through addition of coagulant aids—generally, high molecular weight slightly anionic polyacrylamides.

Water-soluble organic polymers (referred to as polyelectrolytes) are used either as coagulants or flocculants. These may be anionic, cationic or nonionic and are characterized by high viscosity. Thus, a metering pump equipped with a straight-through tubular diaphragm liquid end is normally specified to handle these chemicals. Polymer feed rates in in-line clarifiers are generally lower than those used in conventional clarification, assuming that the raw water characteristics are equal.

The purpose of the polymer is to provide only sufficient neutralization to allow attraction and adsorption of particles through the entire filter bed. Polymer feedrates, therefore, are generally determined through trial-and-error processes. The object is to minimize effluent turbidity and maximize service run length. It may also be necessary to move the polymer injection point to improve turbidity removal, or to use dilution water to properly disperse the polymer.

B. Chlorination

In addition to its use as a bactericide in sewage and industrial wastes, chlorine also destroys sulfides, sulfites and ferrous iron. It is also used in once-through and recirculating water systems to control slime and algae. Chlorine injection improves coagulation in raw water clarification; reduces taste, odor and color-producing materials through oxidation; reduces biological loading; and oxidizes iron and manganese to facilitate removal by settling and filtration.

Process waters may be chlorinated to control bacteria and slime-producing organisms. In the food processing industry, however, dechlorination is required to avoid taste in canned or frozen foods. Fine paper mills have also established critical limits for residual chlorine.

Calcium hypochlorite, a commercial form of chlorine, is widely used to treat industrial waters. Because calcium hypochlorite solutions contain excess alkali, they tend to raise the pH of the water. In hard waters, calcium carbonate precipitation may occur and dictate the addition of a complex phosphate. Conversely, the high alkalinity characteristic improves the quality of soft and highly corrosive waters.

To avoid calcium carbonate precipitation in hard waters, calcium hypochlorite should be fed as a 1 to 2% available chlorine solution. Sodium hypochlorite, which contains much less available chlorine (approximately 15%), can be fed as is or diluted with water. Both forms of this chemical should be handled carefully, and satisfactory corrosion-resistant materials (e.g., polyethylene) should be used for storing and dispensing.

Chlorine may also be used to control plankton and algae in reservoirs. Depending upon the sensitivity and types of organisms present, a 0.2 to 0.3 ppm concentration is generally maintained (concentrations as great as 50 ppm are sometimes required for certain organisms). Here, accurate injection is important due to the toxicity of free chlorine to fish and aquatic life, even at very low concentrations.

The chlorinated cyanurates are principally used in outdoor swimming pools. In many states, automatic chlorine feed systems (including metering pumps) are required. The chlorinated cyanurates are also used in small industrial water treatment applications where ease of handling is required.

Chlorine can assist coagulation by destroying organic contaminants that have dispersing properties. When an inorganic coagulant is used, chlorination prior to coagulant addition establishes the proper pH environment for the primary coagulant. Chlorination prior to primary coagulant feed also reduces the required coagulant dosage. Thus, the most efficient chemical addition sequence is normally:

1. Chlorine.
2. Bentonite.
3. Primary inorganic and/or polymer coagulant.
4. pH adjusting chemicals.
5. Coagulant aid.

Addition of all treatment chemicals (except coagulation aids) should be accompanied by very turbulent mixing of the influent water. High shear mixing is particularly important when cationic coagulant polymers are used. These polymers should be added as far ahead of the clarifier as possible. When adding coagulant aids, however, moderate mixing is recommended to stimulate floc growth; high shear mixing interferes with the polymer's bridging function.

C. Lime Soda Softening

In lime soda softening, lime and soda ash chemically precipitate calcium and magnesium (the hardness producing elements) from raw water. The removal of "hardness" and subsequent use of the softened water minimizes costly problems such as scale and sludge in boilers, condensers, heat exchangers, and other process equipment. In addition, this process is used to soften water for municipal and industrial process use. Lime-soda softening does not reduce hardness to as low a value as zeolite softeners and demineralizers. However, where raw water hardness is relatively low, hot process softeners are sometimes used primarily for silica removal and deaeration, rather than for softening.

Lime soda softening may be effected at normal raw water temperatures (cold process) or at temperatures near the boiling point (hot process). In either instance, accurate addition of chemical reagents to raw water flow is an indispensable factor in efficient processing.

D. Filtration

Filtration plays a key role in many facets of wastewater treatment. Due to the complexity of this subject, the various methods of filtration cannot be detailed here. Metering pumps are mainly used as described in Section V-F.

II. INDUSTRIAL WASTEWATER TREATMENT

A. Refinery and Petrochemical Wastewater

A refinery and petrochemical waste streams are best handled at the source, before they enter the drainage system. This reduces the number of pollutants to be subsequently removed and enables them to be handled at a higher concentration. Additionally, treatment at the source facilitates recovery of many expensive chemicals.

Immiscible liquids and solids, characteristic to these industries, are removed by physical separation techniques such as filtration or centrifuging (preferably at the source, prior to dilution with other streams). Oil is separated from wastewater in American Petroleum Institute (API) separators or parallel plate interceptors.

Solids are separated from water through clarification and settling. Separation of stable emulsions or suspensions can be improved through addition of polyelectrolytes, which causes discrete particles to coagulate. Air flotation may also be used to carry suspended particles to the surface where they are removed by a skimmer. Flocculants and pH control chemicals are sometimes required in air flotation systems; these chemicals are added as shown in Figure 8.1.

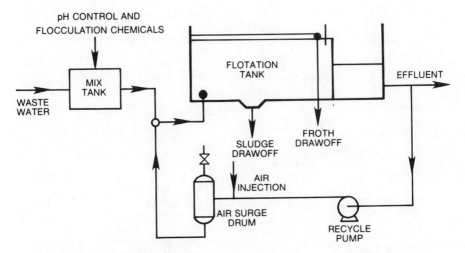

Figure 8.1. Typical air flotation unit. [Reprinted from *Treatment of Industrial Effluents*, John Wiley & Sons, New York (1973).]

Once the wastewater enters the drainage system, cleanup becomes a complex problem. Before a wastewater treatment system can be set up, it is necessary to analyze the effluent to determine the proper physical/chemical treatment methods. Pollutants can enter drainage systems continuously or intermittently—continuous sources are fairly easy to isolate, while intermittent pollutants may be difficult to identify without repeated analyses. Once the pollutants are identified, treatment can be prescribed.

Although wastewater treatment systems are generally custom-engineered for each situation, certain generalizations can be applied to many refineries and petrochemical units. Wastewater consists primarily of process water used in the production of industrial chemical products and cooling system blowdown. This effluent contains organic matter, inorganic salts, suspended solids and other contaminants. The following treatment chemicals are typically used:

acids – H_2SO_4;
alkalies – NaOH;
chlorine – Cl_2;
coagulants – Alum;
coagulant aids – Polymers.

In addition to chemical compatibility, the temperature and pH of the wastewater must also be compatible with the conditions of the natural stream into which it is discharged. The chemical treatment system shown in Figure 8.2 requires a large number of metered chemicals. Many are corrosives; others are

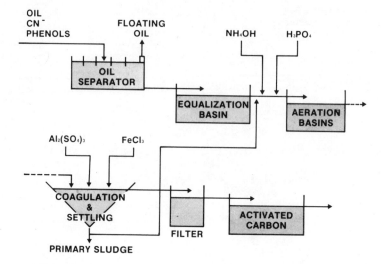

Figure 8.2. Refinery waste.

liquids or slurries. Plant operation is automated through use of metering pump systems—ammonia, phosphoric acid, alum and ferric chloride are added as settling agents.

B. Electric Power Generation Wastewater

Wastewater discharged from electrical power plants consists primarily of boiler blowdown, cooling system blowdown and runoff from fuel storage areas and ash-handling systems. The effluent typically contains suspended solids, metals, salts and other chemicals in solution.

The following chemicals are typically metered into electric power generation wastewater to reduce suspended and dissolved impurities to environmentally acceptable levels.

acids – H_2SO_4;
alkalies – NaOH;
chlorine – Cl_2;
coagulants – alum.

Pretreatment wastes, boiler blowdown and condensate losses are the main sources of wastewater from boiler systems. Ion exchange regenerants and rinse waters are frequently metered into wastewater to control pH. High-pressure

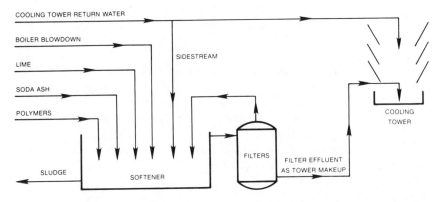

Figure 8.3. Schematic diagram of sidestream treatment. [Reprinted from *Betz Handbook of Industrial Water Conditioning*. Betz Laboratories, Inc., Trevose, PA. 7th ed. (1976).]

boiler blowdown, which is usually low in solids, may be reused as feedwater to lower-pressure boiler systems to reduce the effluent volume.

Cooling towers contribute significant volumes of plant wastewater; thus, high cycles of concentration are desirable. Cascading cooling tower blowdown can also reduce wastewater treatment costs in certain cases.

1) Sidestream Treatment of Cooling Tower and Boiler Blowdown

Some plants use lime-soda softening of cooling tower blowdown to reduce calcium, magnesium, bicarbonate and silica concentrations. By drawing the blowdown from the warm return line, these plants can take advantage of improved hardness and silica removal at the higher temperatures (see Figure 8.3).

An alternate approach is the combination of blowdown and fresh make-up water in the primary lime soda softener. This is generally more acceptable when the make-up water hardness is present in the form of calcium or magnesium bicarbonate. Otherwise, high chloride and sulfate solids tend to create a corrosive environment in the cooling system.

The lime requirements for softening are generally about 30% above theoretical. This is attributed to interfering ions and the use of deposit control agents in the cooling systems competing to hold calcium in solution. Chromate inhibitors return to the system through the softener. Most other inhibitors and deposit control constituents are removed in the softener and their feed must be supplemented.

C. Petroleum and Natural Gas Processing Wastewater

The effluent discharged from a refinery, consists of process water and steam condensate, some of which has come into direct contact with oil and other hydrocarbons. This effluent contains contaminants such as hydrogen sulfide, ammonia, phenols, spent acid and caustic solutions and other organic and inorganic compounds.

Wastewater must be treated before being discharged into the environment. The purpose of treatment is to reduce suspended and dissolved contaminants to environmentally acceptable, safe levels. A typical refinery wastewater treatment schematic is shown in Figure 8.2.

Metering pumps are used to add the following treatment chemicals to wastewater:

alkalies — NaOH;
acids — H_2SO_4;
coagulants;
filter aids;
biological activity enhancers.

D. Pulp, Papermaking, and Textile Wastewater

1) Pulp and Papermaking Wastes

The constituents of paper mill effluent are the raw materials used in the papermaking process. The proportions differ significantly, however. Thus, the more fresh water that is added to the process, the greater the loss of raw materials and the volume of effluent. This makes recycling very attractive, particularly in single-product mills in which continuous production increases the opportunity for recycling.

Because of the high volume of wastewater continually being discharged, pulp and paper mills frequently maintain waste treatment facilities comparable to those found in municipal treatment plants. Wastewater includes such contaminants as color, suspended solids and BOD materials, all of which must be removed. Flocculation and sedimentation are used for removal of suspended matter; trickling filters and activated sludge methods are used for the removal of oxygen-absorbing (BOD) substances in solution.

Preferred flocculants include aluminum sulfate and silica; polyelectrolytes are also popular because they do not contribute to sludge formation. Ammonia and phosphoric acid may also be added after primary settling of large particulate to provide nitrogen and phosphorus sources to maintain organism mass in the larger aeration basins. Later in the process, alum or soda ash slurries may be added—alum as a coagulation and settling agent, soda ash as a settling agent. A typical water treatment schematic is shown in Figure 8.4.

b) Textile Wastes

Cleanup is very similar to that used for the pulp and paper process, except sanitary waste is sometimes substituted for ammonia and phosphoric acid as a nitrogen and phosphorous source. Use of sanitary waste requires addition of chlorine as a final step.

E. Food and Beverage Processing Wastewater

The food and beverage processing industry has many segments, but the major water-using segments are: sugar cane/beet processing, fruit and vegetable processing, grain processing, meat and poultry, fats and oils, dairy products, and beverage manufacturing.

Wastewater from food and beverage processing is composed of water discharged from cooling, boiler and processing operations. It is used in washing raw food materials or process equipment, dissolving or extracting, and conveying products from one process area to another. Depending on the type of food processed, effluent may contain such suspended and dissolved materials as soil, vegetable peels, organic solids, spent acid and alkaline solutions, fats, greases and oils.

Before discharge into the environment, wastewater must be chemically and physically treated. Treatment is necessary to reduce suspended and dissolved pollutants to environmentally acceptable levels. Before discharge into streams or as groundwater, wastewater must be as similar as possible to natural water in terms of temperature, chemical composition, color and pH.

Metering pumps are used to add the following chemicals to wastewater discharged from food and beverage processing plants:

coagulant aids;
acids — H_2SO_4;
alkalies — NaOH;
filter aids.

Although food and beverage processing wastewater is generally more difficult to treat than domestic sewage, many of the same biological treatment

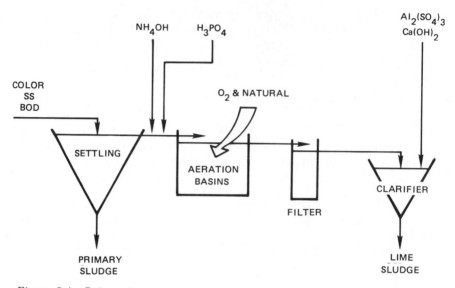

Figure 8.4. Pulp and paper waste.

Figure 8.5. Pulp and paper wastewater treatment.

methods may be applied. These include activated sludge and biological filters. For effective biooxidation of organic matter, there is an optimum ratio of BOD: nitrogen:phosphorous of approximately 100:6:1.

Wastewater deficient in nutrients (e.g., brewing and canning) may require supplementing prior to treatment to help overcome the problem of "bulking" in the activated-sludge process.

The application of high-strength wastes to conventional biological filters may cause ponding, which is overcome by the use of alternating double filtration. The problems of high-strength organic wastes are also solved by the use of filters with plastic media, especially when used as first-stage roughing filters. Research is also being carried out on the use of anaerobic systems for treating high-strength wastes, using either modified digesters or anaerobic filters.

Further simplifications may be used, especially in rural areas where land and odor problems are not limiting, including spray irrigation and lagooning.

1) Pretreatment

Biological (secondary) treatment generally follows a pretreatment process for economic and operational reasons; screening, sedimentation, flocculation, air flotation, filtration and centrifugal separation may be used.

2) Physical and Chemical Processes

The use of membrane techniques such as reverse osmosis and ultrafiltration is becoming fairly widespread, and as reuse becomes essential, evaporation, adsorption and protein recovery are finding increased application in many food industries.

3) Sludge Treatment

The treatment of solids, both those separated from the raw waste and those generated by biological processes, is possibly the most difficult aspect of waste treatment, and may result in up to half of the capital and operating costs. Most of those systems used in municipal sewage-treatment plants may be used; system selection depends upon the nature of the industry and the treatment process.

F. Wastewater from the Metalworking Industries

1) Iron and Steel Waste

A variety of contaminants, including suspended solids, phenols, cyanides, sulfides and ferrous compounds, must be removed from the effluent stream. After

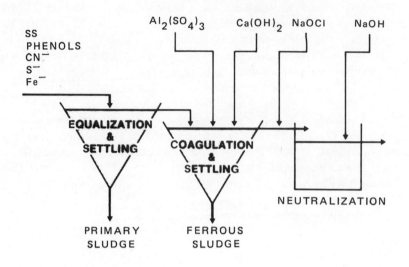

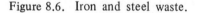

Figure 8.6. Iron and steel waste.

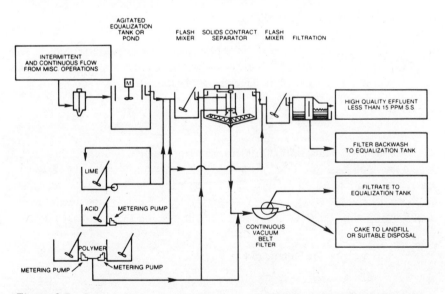

Figure 8.7. Low volume, metal cleaning, and area runoff wastes. [Reprinted from *Betz Handbook of Industrial Water Conditioning*. Betz Laboratories, Inc., Trevose, PA, 7th ed. (1976).]

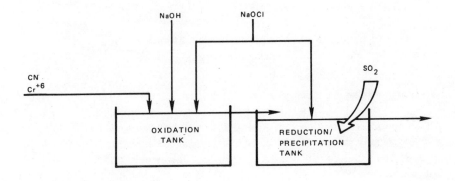

Figure 8.8. Plating waste.

primary settling, soda ash and alum are dispensed for coagulation and settling. Other chemicals are metered into the effluent to remove toxic contaminants. For example, sodium hydroxide may be injected for cyanide removal and sodium hypochlorite added to kill bacteria. Typical water treatment isometrics are shown in Figures 8.6 and 8.7.

2) Plating

Effluents containing various highly toxic compounds are yielded. An example of the treatment steps used in removal of cyanide and hexavalent chromium is shown in Figure 8.8.

Although the flow diagram is simple, plating waste is the most difficult to treat. NaOH is used for cyanide removal; SO_2 is used to reduce the chromate to a trivalent compound that is then precipitated with such metered chemicals as sodium hydroxide.

Depending on the other contaminants, chemicals are injected by batch or continuous treatment systems. Continuous systems handle large volumes of wastewater that is fairly consistent in composition; batch systems are employed for small or intermittent volumes or if flow rates and chemical compositions vary substantially.

III. POTABLE WATER TREATMENT

Treatment of potable water involves elimination of inorganic and organic chemicals, turbidity, and microbiological contaminants.

A. Inorganic Chemicals

This is the most difficult group to deal with since conventional treatment techniques generally are not effective. Therefore, first consideration should be given to alternative sources and blending of sources. Fortunately, the inorganics usually are not present in high concentrations. They are generally found naturally only in certain ground waters or only when industrial contamination is present. However, lead and cadmium contamination can occur as a result of an aggressive water attacking plumbing systems where lead or galvanized piping is still in use. This problem can best be handled by stabilizing the water through chemical addition.

When treatment must be considered, choices are somewhat limited, as summarized in Table 8.3.

B. Organic Chemicals

The widespread use of certain pesticides has increased the potential for contamination of drinking water. Generally, this is only a concern with surface water due to the influence of runoff. The most effective methods for removing these compounds are to use powdered (PAC) or granular activated carbon (GAC). Pesticides account for only a few of the more than 700 organic compounds identified in drinking water supplies throughout the United States. These compounds result from such sources as industrial and municipal discharges, urban and rural runoff and natural decomposition of vegetative and animal matter. Compositions and concentrations vary from virtually none in protected ground water supplies to substantial levels in many surface waters.

C. Turbidity

Turbidity is best dealt with by direct filtration or more complete treatment systems.

D. Microbiological Contaminants

Disinfection is used to control microbiological contaminants. The choice of disinfectant depends on a number of factors including raw water quality and

TABLE 8.3

Treatment Methods for Inorganic Contaminant Removal

Contaminant	Methods	% Removal
Arsenic		
As^{3+}	Oxidation to As^{+5} required	> 90
As^{5+}	Ferric sulfate coagulation, pH 6–8	> 90
	Alum coagulation, pH 6–7	> 90
	Lime softening, pH 11	> 90
Barium	Lime softening, pH 10–11	> 80
	Ion exchange	> 90
Cadmium*	Ferric sulfate coagulation, > pH 8	> 90
	Lime softening > pH 8.5	> 95
Chromium*		
Cr^{3+}	Ferric sulfate coagulation, pH 6–9	> 95
	Alum coagulation, pH 7–9	> 90
	Lime softening, pH > 10.5	> 95
Cr^{6+}	Ferrous sulfate coagulation, pH 6.5–9 (pH may have to be adjusted after coagulation to allow reduction to Cr^{+3})	> 95
Fluoride	Ion exchange with activated alumina or bone char media (limited full-scale experience, pH slightly above 7	> 90
Lead*	Ferric sulfate coagulation, pH 6–9	> 95
	Alum coagulation, pH 6–9	> 95
	Lime softening, pH 7–8.5	> 95
Mercury*		
Inorganic	Ferric sulfate coagulation, pH 7–8	> 60
Organic	Granular activated carbon	> 90
Nitrate	Ion exchange (limited full-scale experience)	> 90
Selenium*		
Se^{4+}	Ferric sulfate coagulation, pH 6–7	70–80
	Ion exchange	> 90
	Reverse osmosis	> 90
Se^{6+}	Ion exchange	> 90
	Reverse osmosis	> 90
Silver*	Ferric sulfate coagulation, pH 7–9	70–80
	Alum coagulation, pH 6–8	70–80
	Lime softening, pH 7–9	70–90

*No full-scale experience.

Reprinted from *Water Treatment Plant Design*, Ann Arbor Science Publishers, Inc., Ann Arbor, MI, Second printing (1979).

amount of water to be treated. Although chlorine is preferred, if the raw water contains enough organic compounds to result in formation of trihalomethanes, a disinfectant other than chlorine may be chosen. If this is done, however, an evaluation is necessary to define the proper operating parameters and ensure proper disinfection.

1) Chlorination of Potable Water

Chlorine is metered into public water supplies to control taste and odor and to disinfect the water; to oxidize iron, manganese, and hydrogen sulfide; and to oxidize, decolor, and destroy organic matter. Chlorine dioxide, which may also be used, is noted for its ability to destroy phenols without creating the taste problems of chlorinated phenols.

9

ENERGY CONSIDERATIONS

Electric energy will probably always represent an appreciable fluid handling cost component. For this reason, pump electrical consumption should be minimized wherever possible.

This chapter addresses several energy-saving methods that apply to metering pumps. Many of the tips given here apply to other pump families as well.

I. EQUIPMENT SELECTION

As energy costs continue to escalate, pumps should be closely matched with service conditions, notably capacity. Oversizing, a safety factor traditionally employed by systems engineers, is economically unattractive for obvious reasons.

Operating pumps close to the maximum capacity limit not only requires additional engineering effort and manufacturer consultation, but also a rugged, reliable design that can operate at or near peak loads for considerable periods of time. This is particularly true in the continuous operations commonly found in the chemical processing industry (CPI). A proven design should be used to ensure reliable service under increasingly demanding conditions.

In the past, pumps were typically sized to run at about 70% of anticipated system capacity. Capacities were determined on a theoretical basis and then increased or decreased after actual start-up, with changes usually minor.

In applications in which the designer can draw upon previous experience, oversizing is not necessary. Wastewater treatment is one such application. But in the CPI, where few systems are alike, it is often difficult to refer to previous design experience when specifying the pump.

In the future, much closer attention will be paid to system design in the early stages, particularly in terms of sizing motors. The following example illustrates this point.

A metering pump that must deliver 40 gph @ 360 psi discharge pressure and then is oversized 30% is actually capable of delivering 58 gph @ 360 psi. Such a pump requires a 1-hp motor. However, if the pump were exactly sized, a 3/4-hp motor could be substituted to provide the required flow. This represents a long-term savings in "motor inertia" power requirements.

Metering pump motors are sized as follows:

$$HP = \frac{30.8 \times 10^{-6} \, pQ}{\mu}$$

where p = total head pressure delivered by pump, in psi
Q = capacity of water-like fluid delivered, in gal./hr.
μ = pump efficiency, expressed as a decimal

II. MULTIPLEXING

In addition to proper sizing of electric motors, multiplexing can also save energy (Figures 9.1-9.3). Due to the sinusoidal characteristic of the flow curve, the pump motor handles the peak load while "resting" between peaks. By duplexing pumps (driving two plungers with the same motor) and synchronizing the units so the sinusoidal output of each is 180° out-of-phase, twice the output is

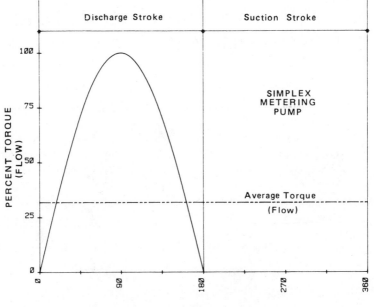

Figure 9.1. Torque (and flow) characteristic of a simplex metering pump varies sinusoidally, with peak load occurring when the plunger has traversed through 50% of the discharge stroke.

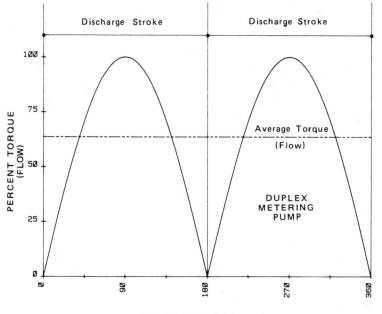

Figure 9.2. Use of two plungers (duplex arrangement) operating 180° out-of-phase permits a doubling of flow without affecting maximum torque required.

theoretically possible from the same motor. A duplex pump is shown in Figure 9.4. When three pumps are synchronized (120° plunger phasing) and employ a common driver, a triplex arrangement results (Figure 9.5) with the peak load occurring just slightly below that of a simplex or duplex pump. This analysis ignores inertial and frictional loads which should be taken into account under high thrust load conditions, and when using large-diameter placked plunger liquid ends.

When considering a duplex or triplex arrangement, the user must remember that the additional cost of the suction and discharge line manifolds will increase the initial cost. Rapid payback will often justify such an arrangement, however.

III. TWO OR MORE PUMPS

Two or more pumps of different sizes with separate motors should be employed where large changes in capacity or turndowns are expected.

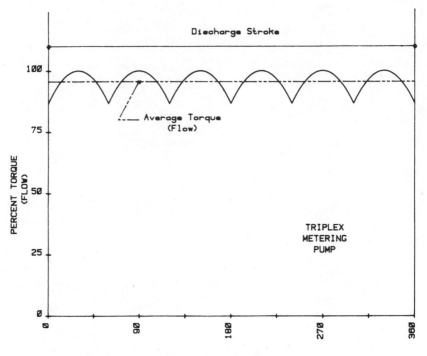

Figure 9.3. Operation of three plungers (triplex arrangement) 120° out-of-phase yields a peak torque equivalent to simplex and duplex arrangements.

Figure 9.4. Duplex metering pump arrangement .

Figure 9.5. Triplex metering pump arrangement.

In a power plant, for example, the smaller pump should be run during off-peak hours in place of the larger pump. A centrifugal pump with most of its capacity diverted through a bypass valve during off-peak hours still consumes the same amount of current as when it is producing at 100%. If a smaller pump were connected in parallel, its smaller motor would consume much less energy when operated in place of the larger unit.

The same holds true for metering pumps. Although they feature a 10:1 turn-down ratio, which means capacity can be varied from 100% to 10%, the amount of current drawn does not drop proportionally due to inertial and frictional loads.

Two-speed motors, or two separate drive motors, are the answer to some pump energy consumption problems. The two motors should be sized to provide sufficient difference in capacity and head to meet minimum and maximum process requirements.

IV. VARIABLE-SPEED DRIVES

Electric variable-speed drives are supplanting high-loss control valves, mechanical speed reducers and automatic capacity controls due to product improvements arising from advancements in electronic technology. Selection of the proper driver for a pump, however, must take into account cost and complexity of the system, motor characteristics desired, method of speed control and operating range.

Solid state drives for DC motors may use silicon controlled rectifiers for half- or full-wave rectification and adjustment of voltage, or they use a transistor amplifier to generate the variable DC. A commonly used DC variable speed controller contains a full wave bridge rectifier, an SCR, an SCR triggering circuit and a filter capacitor. Where close speed regulation is essential, a feedback loop may be used to compensate for changes in motor load, line voltage, ambient temperature and other disturbances.

Because AC motor speeds are dependent on line frequency, control of speed by varying the frequency of the power to the motor is a natural choice. Latest state-of-the-art drives employ this technique, along with variable voltage supply. Less complex drives can be derived by using SCR's to control power to the motor at fixed input frequency.

Practically any size and type of adjustable speed drive is available from the various manufacturers. In addition to standard designs, drives may be custom designed for special applications.

Variable speed drives are already widely used in water and wastewater treatment applications, boiler feed service and circulating water pumps in power stations. Metering pumps used in continuous processes and metering of chemicals are also frequently equipped with variable speed drives. Used in these and other applications, variable speed drives offer average regulation accuracy of 3%, higher starting torques at lower speed settings and speed ranges of 30:1 and better.

V. ENERGY EFFICIENT MOTORS

Although the demand for an energy-efficient (EE) motor has always existed, attention by the user has been in the past devoted to bottom line cost and such factors as power rating, speed reliability, NEMA rating, and duty cycle. But dramatic changes in the energy situation have suddenly caused a previously minor consideration—energy consumption—to spring to the forefront.

It is not surprising that the Department of Energy (DOE) has selected electric motors as one of the top 11 "exceptional near-term opportunities" for achieving energy conservation. DOE studies have indicated that electric motors used 75% of the 600 billion kWh of the electricity consumed by industry each year, or 28% of all electric power generated in the U.S.

As a result of this new impetus, EE motors have evolved. These motors offer up to 10% greater efficiency and up to 20% higher power factor. As a result of these improvements, other benefits are derived, such as lower operating temperature, increased life, reduced sensitivity to high and low voltages, high overload capability and the ability to better overcome inertia loads.

Although cost is typically 5% to 25% more than standard motors, payback can be determined based on local utility rate structures and power factor charges. The greatest potential for substantial energy savings is generally offered by AC polyphase motors in the 1- to 125-hp range, with models available in all standard NEMA configurations.

EE motors produced by several manufacturers are designed for optimum electrical and mechanical efficiency. Longer stator/rotor cores are used to reduce magnetic density and cut power consumption. More copper is used in windings in place of aluminum, and the air gap between them is optimized. One manufacturer says its EE motors dissipate up to 27% less power in the form of heat and may run up to 60% cooler. In theory, every 10% reduction in operating temperature theoretically doubles insulation life.

Some metering pumps employ air-operated cylinders as a mode of power. Although a very important factor when pumping is required on a stroke basis in closed loop systems, the pneumatic power source remains less efficient than electric power.

Pneumatic power remains a must, however, when a metering pump is used in a very hazardous atmosphere or in a location where power is not readily available.

10

MAINTENANCE GUIDELINES

Metering pumps are precision instruments in that they are designed to deliver very precisely controlled volumes of liquids per unit of time with adjustable output. Throughout the pumps' capacity range, delivered volume is generally controllable to within ±1% of setting.

As precision instruments, metering pumps should not be treated as "just another pump." The same careful attention should be given to their installation as is given to other precision instruments to assure longer, more accurate and more dependable service.

I. GENERAL PREVENTIVE MAINTENANCE

Newly installed pumps should be checked at least once every 24 hours until proper maintenance schedules, based on actual operating experience, have been established.

A. Spare Parts

Order spares early. Most manufacturers supply a list of spare parts showing which should be stocked and the minimum recommended quantity of each. Ideally, parts should be ordered along with the pump or, at the latest, when the pump is put into service. This is particularly important when the pump is an integral part of a continuous process in which downtime is costly.

Establish a regular preventive maintenance schedule. It is impossible to overemphasize the importance of such a schedule. Whenever possible, parts should be replaced before they fail. Parts likely to wear include rod bushings, packings, plungers, ball checks, ball seats and diaphragms. The life expectancies of these and other parts can be determined by operating experience.

B. Lubrication

Set up a firm lubrication schedule. If a pump has been properly installed, the only additional maintenance that should be required is routine lubrication.

TABLE 10.1

Packing Lubricants

Lubricant Type	Temperature Range, °F	Type of Service
Highly purified petroleum (hydrocarbon oils and waxes)	20 to 125	Acids, alcohols, alkalies, aqueous solutions, glycerin, dyes, food and pharmaceutical (as determined by user).
High temperature lubricating composition of selected liquid and solid polymeric hydrocarbons (non-soap, non-mineral thickened)	0 to 650	Acids, alkalies, alcohols, amines, hot aqueous solutions and glycols.
Non-soap thickened vegetable oil composition	−20 to 500	General purpose lubricant for hydrocarbon liquids.
Synthetic composition containing non-soap thickened liquid polyols	−40 to 160	Hydrocarbon and aromatic solvents.

Proper and adequate lubrication on a regular schedule will greatly extend pump life.

Lubricate all moving and sliding parts once every eight hours if the pump is under continuous full load. If the pump is operated intermittently, it should be lubricated in proportion to operating time.

When packed plunger pumps are employed, lubricate the plunger packing daily. The lubricant must be compatible with the liquid being pumped. Table 10.1 lists the types of lubricants that should be used with commonly metered fluids. Do not substitute any other kind of lubricant without consulting the manufacturer. Pumps with plastic liquid ends may not require periodic lubrication of packing.

Check gear drive oil monthly. If level is low, add oil as required. Metering pumps that have bronze worm gears should be filled with AGMA 8 Compounded or a similar lubricant containing inactive extreme pressure (EP) additives. Change gear drive oil every six months, or after 2,500 hours of operation, whichever comes first. At the same time, clean the magnetic filter (if provided) below the crosshead chamber.

Lubricate motors yearly or more often if the motor manufacturer's instructions so indicate.

II. PACKED PLUNGER PUMPS

When packed plunger pumps are employed, check the condition of the plunger on a regular schedule, especially when excessive packing wear occurs. Packings will not last very long if used with a scored plunger. A lightly scored plunger can be polished on a lathe; a badly scored plunger should be replaced.

Check the fit of the crosshead-plunger assembly into the pump frame on a regular schedule. This can be done when replacing packing. If the fit is too loose, the weight of the plunger is on the packing instead of on the crosshead. This condition results in rapid deterioration of the packing and allows the plunger to drag on the lantern ring or throat of the stuffing box. Scoring of the plunger results. If there is evidence of wear on the plunger guiding surfaces (crosshead and/or pump frame), they should be machined or replaced.

A. Packings

Pumps with packed plunger type liquid ends require periodic replacement of packing. Generally, packings depend for their proper functioning on compression in the stuffing box. If too tightly compressed, packing will overheat and may disintegrate rapidly. Excessively tight packing may score the plunger, necessitating replacement of the plunger itself.

Use the correct packing for liquid being pumped. See Figure 1.1 for recommendations concerning the best type of packing for pumping a variety of liquids.

Do not substitute an automatic packing set for a compression packing set or vice versa. Two general types of packing are used in metering pumps:

1. Automatic, self-sealing "V" or Chevron types.
2. Compression types.

One type of automatic packing can normally be substituted for another automatic packing without causing trouble, but usually an automatic packing set should not be replaced with a compression set or vice versa, unless there is a good reason to make the change.

Automatic "V" or Chevron-shaped packings depend for their operation on *hydraulic pressures* in the displacement chamber which act on their special shapes. On the discharge stroke of the pump, sealing edges expand into close contact with the plunger and the inside surface of the stuffing box, thereby effecting a seal. Automatic packing must never be operated so tightly that it cannot properly flex. It is seldom necessary to operate automatic packing more than "fingertight," except under extreme pressure conditions.

Compression type packings are of square or round cross-section and must be firmly compressed in the stuffing box. They are made of many different compositions, each specifically designed to handle a different group of chemicals. These types of packings depend entirely upon *gland pressure* to properly effect the liquid seal between plunger/stuffing box and packing sealing surfaces.

B. Corrective Maintenance

A proper preventive maintenance schedule including regular lubrication and routine replacement of worn parts before they fail in service will greatly reduce the likelihood of emergency breakdowns; however, emergencies do happen. In the event of a breakdown or failure of the pump, the trouble-shooting chart (Appendix B) may prove helpful.

Replacement of packing is detailed in the following section: (Refer to Figure 10.1)

1. Thoroughly clean the stuffing box of old packing and grease.
2. Check the plunger for "droop" (poor fit of the crosshead caused by wear).

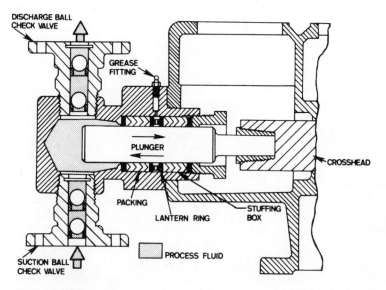

Figure 10.1. Packing replacement.

3. Check the plunger for scoring or pitting by corrosion. Repolish or replace as necessary.
4. Fill the stuffing box with new packing to within about ¼ inch of the top.
   - Cover each piece of packing thoroughly with the lubricant recommended by the pump manufacturer.
   - Place one packing section at a time in the stuffing box. If adapters are required, be sure to locate them properly at the bottom and top of the packing and on both sides of the lantern ring (if a lantern ring is used).
   - Tamp each piece of packing in place before adding the next piece. If the rings are split, stagger the joints so they are not in line.
   - Locate the lantern ring carefully, preferably with the end toward the pressure side just under the lubrication fitting. It will move up as the packing is "set" and adjusted. Filling the stuffing box to within ¼ inch of the top allows the gland to enter sufficiently to provide proper guidance.
5. Place the gland in position. Pull up evenly on the gland until it is under heavy compression. Hold this pressure on the packing for several minutes to allow it to flow into position in the stuffing box.
6. Follow packing break-in instructions (see Installation and Start-Up Guidelines).

Many new packings, including plastic packings with high corrosion resistance, have been introduced in recent years. Although these packings are considerably more expensive than conventional packings, they possess an almost complete chemical inertness which makes it economical to use them for many applications. Special instructions are usually provided by the manufacturers of such packings, and these instructions should be followed closely.

III. DIAPHRAGM PUMPS

Monitor and maintain the volume of hydraulic fluid in the displacement chamber. The plunger reciprocates in a chamber filled with hydraulic fluid. The hydraulic fluid lubricates all moving parts.

The plunger displaces the hydraulic fluid, which actuates the diaphragm to create pumping action. The diaphragm is hydraulically balanced by the hydraulic fluid in the displacement chamber on one side and the liquid being pumped in the head, which is on the other side of the diaphragm. Double ball check valve cartridge assemblies on both suction and discharge ports of the diaphragm head

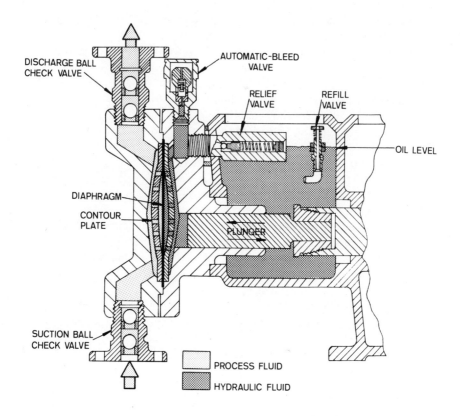

Figure 10.2.   Three-way valve system.

ensure positive repetitive accuracy of the liquid being metered. Repetitive metering accuracy can be obtained only if the volume of hydraulic fluid in the displacement chamber is maintained.

A 3-valve system is often used on the displacement chamber (Figure 10.2):
1) a refill valve replenishes any hydraulic fluid that leaks out of the chamber,
2) an air-bleed valve removes air or gas bubbles from the hydraulic fluid, and
3) a relief valve provides over-pressure protection for both hydraulic and process
fluids.

If pump has two diaphragms (disc and tubular), the two diaphragms
should be kept in synchronization to ensure that the tubular diaphragm is in its
fully relaxed position at the same time as the disc diaphragm is against the contour on the drive (hydraulic fluid) side. This allows full flexing travel of both
diaphragms simultaneously with minimum stress on the tubular diaphragm, thus

eliminating complete collapse at full discharge stroke position. Pumps are normally shipped with the intermediate chamber primed and sealed with a 50% mixture of glycol and water and the diaphragms synchronized; therefore, synchronization is *not* necessary until such time as it becomes necessary to replenish or replace the liquid in the intermediate chamber with another fluid. CAUTION: *IF PUMP HAS BEEN USED FOR PUMPING HAZARDOUS LIQUIDS, FLUSHING IS NECESSARY TO PROTECT PERSONNEL.* See manufacturer's instructions.

MISCELLANEOUS

Always consult the pump manufacturer before using a pump for any process fluid other than that for which it was purchased. When communicating with the manufacturer, be sure to supply all pertinent details including pump serial number. Most manufacturers keep detailed records, by serial number, of all pumps shipped. Exact engineering details of any pump in the field can be determined easily by reference to the serial number.

When contacting manufacturers, specify the liquid to be pumped, its temperature and other known characteristics. A rough sketch of your installation, showing liquid levels in tanks and the dimensions and location of both suction and discharge piping, is also helpful.

APPENDIX A

SHORT GUIDE TO SELECTING METERING PUMPS

This guide is not designed to be the sole tool for selecting one type of metering pump over another. It is no more than a starting point—a very general set of rules to which there are exceptions. For additional information, see Chapter 3, Selecting Metering Pumps, and Chapter 6, Installation.

I.   PACKED PLUNGER LIQUID END

Recommended for:

1.   Liquids with temperatures that exceed 250°F.
2.   Non-corrosive or mildly corrosive liquids with high viscosities.
3.   Solvents or fluids with low vapor pressures. Packed plunger pumps require only enough suction pressure at the liquid end to keep the liquid from vaporizing in the liquid end, plus a 3 psia safety margin.
4.   High suction pressures. Maximum suction pressure limits are greater for packed plunger pumps than for diaphragm pumps.
5.   Applications requiring very high discharge pressures.

Not Recomended for:

6.   For extremely corrosive or hazardous liquids, due to packing leakage. If extremely corrosive fluids are allowed to accumulate in the pump catchall chamber, damage to the housing can result. Corrosive fluids can also damage the crosshead and limit pin. Even when funneled off, splashing can cause corrosive action. When the plunger is moving rapidly or the discharge pressure is very high, the packing leakage problem can be aggravated.
7.   For applications where routine maintenance (packing replacement) is not possible.
8.   For fluids that must be isolated from the pump drive oil.

II. DISC DIAPHRAGM LIQUID END

Recommended for:

1. Extremely corrosive liquids, due to leak-proof features.
2. Applications where low maintenance is required.

Not Recommended for:

3. For high viscosity fluids, because the contour plate holes restrict the flow of viscous fluids.
4. For very abrasive slurries, because particulate may accumulate between the contour plate and the diaphragm.
5. For liquids with temperatures that exceed 250°F.
6. When solvents or fluids having low vapor pressures are used, and suction pressure is so low that the liquid vapor pressure is reached due to the vacuum created by the pump suction stroke. A minimum of 9 psia suction pressure, plus 3 psia safety margin, is needed for proper operation of the hydraulic refill valve.
7. Applications requiring very high discharge pressure.
8. When suction pressure exceeds the limit specified in the pump instruction manual.
9. For fluid that must be isolated from the pump drive oil in the event of diaphragm failure.

III. DOUBLE DIAPHRAGM LIQUID END

Recommended for:

1. Extremely corrosive or non-corrosive liquids with high viscosities.
2. Abrasive slurries with high or low viscosities.
3. Applications where low maintenance is required.
4. Fluids that must be isolated from the pump drive oil in the event of process diaphragm failure.
5. Applications requiring a positive warning of diaphragm failure.

Not Recommended for:

6. For liquids with temperatures that exceed 175°F (elastomer diaphragm) and 250°F (Teflon diaphragm).

7.   When solvents or fluids having low vapor pressures are used, and suction pressure is so low that the liquid vapor pressure is reached due to the vacuum created by the pump suction stroke. A minimum of 9 psia suction pressure, plus 3 psia safety margin is needed for proper operation of the hydraulic refill valve.

8.   Applications requiring very high discharge pressures.

9.   When suction pressures exceed the limits specified in the pump instruction manual.

APPENDIX B

TYPICAL TROUBLE-SHOOTING CHART FOR METERING PUMPS

Trouble	Cause	Remedy
Pump fails to operate	Blown fuse	Replace blown fuse
	Open thermal overload device in starter	Reset
	Low liquid level (where low-level cutoff is used)	Fill tank
	Broken wire	Locate and repair
	Low voltage	Determine reason (wiring gauge may be inadequate)
	Discharge line blocked	Remove block
	Liquid "frozen" in pump	Thaw out
Pump fails to deliver rated capacity	Starved suction	Replace suction piping with larger size or increase suction head
	Leaky suction piping	Repair or replace defective piping
	Excessive suction lift	Rearrange equipment location to reduce suction lift
	Liquid too close to boiling point	Lower temperature or increase suction pressure
	Capacity adjustment incorrectly set	Adjust properly

Trouble	Cause	Remedy
	Leaky packing*	Adjust or replace packing
	Pump operating at correct speed	Check line voltage and frequency against motor nameplate
	Worn or dirty valve seats, or both	Clean or replace
	Viscosity of liquid too high	1) Reduce viscosity by heating or other means 2) Increase size of suction piping 3) Increase suction pressure
	Air in hydraulic system	Stop pump at end of suction stroke; remove air bleed valve, allow air to escape, fill displacement chamber with hydraulic oil, and reinstall valve
Pump delivers erratically.	Leaky suction line	Repair or replace piping
	Leaky packing*	Repair or replace packing
	Worn or dirty valve seats, or both	Clean or replace
	Excessive excursion of ball valves from seats	Limit excursion to manufacturer's tolerance
	Insufficient suction pressure	Increase suction pressure: 1) Raise tank level 2) Pressurize suction tank
	Liquid too close to boiling	Reduce temperature or raise suction pressure
	Leaky safety valve	Repair or replace safety valve

*Packed plunger pumps only.

Trouble	Cause	Remedy
Motor overheats.	Power supply does not match motor characteristics	Check power supply against motor nameplate data
	Insufficient quantity or improper type of lubricant in gear case	Check level and type of lubricant; replace if doubt exists
	Overload caused by operating pump beyond rated capacity	Check operating conditions against pump manufacturer's specifications
	Packing too tight or improperly lubricated*	Readjust packing and lubricate if necessary
	Pump operating mechanism improperly lubricated	Check all lubricating points
	Mechanical misalignment	Check alignment of all working parts
Noisy operation:		
1) In pump.	Pump valves	Valves must move to open and close, and they will make a clicking noise as they operate. In larger sized pumps, the ball valves may roll in the cages causing rattling noises, these noises are sometimes amplified by natural resonances in the piping system. They are usually indications of normal valve functioning
2) In gear.	Excessive backlash in gears	Replace gears
	End play in high speed shaft	Readjust per instructions

*Packed plunger pumps only

Trouble	Cause	Remedy
	Worn bearings	Replace bearings
	Lack or wrong type of lubricant	Drain and refill gearcase with proper quantity and type of lubricant
Leakage of oil around pump shaft.	Oil seal damaged or worn	Disassemble pump; replace seal and reassemble per assembly instructions
Repeated knocking noise with each stroke.	A slight knock is permissible; unusually loud knocking is due to excessive wear of worm and worm gear	Disassemble pump and replace worm and worm gear; reassemble per pump assembly instructions
Repeated knocking noise with each stroke.	Cavitation (pump starved)	1) Examine suction piping for clogged strainer or obstruction in piping or suction check valve 2) NPSH satisfactory?

APPENDIX C

CALCULATING NPSH

NPSH (net positive suction head) is the hydraulic pressure available above the vapor pressure for feeding liquid into the pump suction port. This pressure must be high enough to overcome those acceleration and velocity losses, associated with a given metering pump system, that can produce cavitation.

Because acceleration loss occurs during a different portion of the suction stroke than velocity loss, each must be analyzed separately. The one which yields the lowest pressure within the liquid end must be considered the limiting factor. In the case of viscous liquids, velocity losses predominate and standard pressure drop equations apply. For water-like liquids (less than 50 centipoise), acceleration losses are higher and satisfactory operation can only be obtained when $P_{acc} \leqslant P_a \pm P_h - P_v$,

$$\text{where } P_{acc} = \frac{(\text{S.G.}) \; L_p \; T \, N^2 \, D^2}{1.36 \quad 10^5 \quad D_p{}^2}$$

$$\text{or } P_{acc} = \frac{(\text{S.G.}) \, Q \, N \, L_p}{2.76 \quad 10^4 \quad D_p{}^2}$$

P_{acc} = Acceleration pressure loss (psi)
S.G. = Liquid specific gravity
L_p = Actual pipe length (ft.)
T = Plunger travel (in.)
N = Pump speed (strokes/minute)
D = Plunger diameter (in.)
D_p = Pipe I.D. (in.)
Q = Flow rate (gph)
P_a = Pressure above liquid (psia)
P_h = Liquid head (psig)
P_v = Liquid vapor pressure (psia)

Liquid head should be based upon the lowest solution level in the supply tank, with vertical distance measured from plunger center line (see Figure C.1).

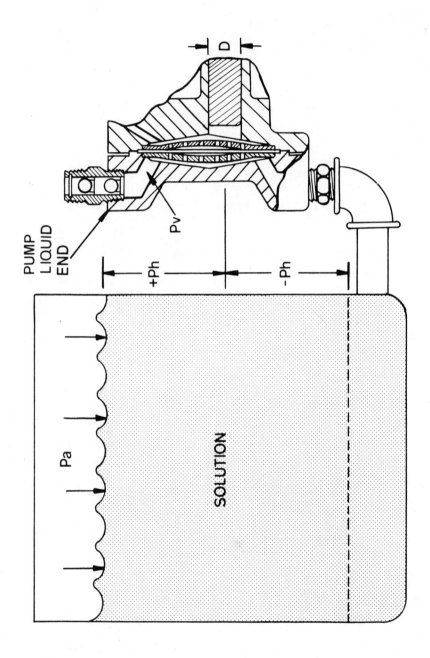

Figure C.1.

APPENDIX D

COMMON IMPURITIES FOUND IN FRESH WATER

Constituent	Difficulties Caused	Means of Treatment
Turbidity	Imparts unsightly appearance to water. Deposits in water lines, process equipment, etc. Interferes with most process uses.	Coagulation, settling and filtration.
Color	May cause foaming in boilers. Hinders precipitation methods such as iron removal and softening. Can stain product in process use.	Coagulation and filtration. Chlorination. Adsorption by activated carbon.
Hardness	Chief source of scale in heat exchange equipment, boilers, pipe lines, etc. Forms curds with soap, interferes with dyeing, etc.	Softening. Demineralization. Internal boiler water treatment. Surface active agents.
Alkalinity	Foaming and carryover of solids with steam. Embrittlement of boiler steel. Bicarbonate and carbonate produce CO_2 in steam, a source of corrosion in condensate lines.	Lime and lime-soda softening. Acid treatment. Hydrogen zeolite softening. Demineralization by anion exchange.
Free Mineral Acid	Corrosion	Neutralization with alkalies.
Carbon Dioxide	Corrosion in water lines and particularly steam and condensate lines.	Aeration. Deaeration. Neutralization with alkalies.
pH	pH varies according to acidic or alkaline solids in water. Most natural waters have a pH of 6.0–8.0.	pH can be increased by alkalies and decreased by acids.

Constituent	Difficulties Caused	Means of Treatment
Sulfate	Adds to solids content of water, but in itself, is not usually significant. Combines with calcium to form calcium sulfate scale.	Demineralization
Chloride	Adds to solids content and increases corrosive character of water.	Demineralization
Nitrate	Adds to solids content, but is not usually significant industrially. High concentrations cause methemoglobinemia in infants. Useful for control of boiler metal embrittlement.	Demineralization
Fluoride	Cause of mottled enamel in teeth. Also used for control of dental decay. Not usually significant industrially.	Adsorption with magnesium hydroxide, calcium phosphate, or bone black. Alum coagulation.
Silica	Scale in boilers and cooling water systems. Insoluble turbine blade deposits due to silica vaporization.	Hot process removal with magnesium salts. Adsorption by highly basic anion exchange resins, in conjunction with demineralization.
Iron	Discolors water on precipitation. Source of deposits in water lines, boilers, etc. Interferes with dyeing, tanning, paper making, etc.	Aeration. Coagulation and filtration. Lime softening. Cation exchange. Contact filtration. Surface-active agents for iron retention.
Manganese	Same as iron.	Same as iron.
Oxygen	Corrosion of water lines, heat exchange equipment, boilers, return lines, etc.	Deaeration. Sodium sulfite. Corrosion inhibitors.
Hydrogen	Cause of "rotten egg" odor. Corrosion.	Aeration. Chlorination. Highly basic anion exchange.

Constituent	Difficulties Caused	Means of Treatment
Ammonia	Corrosion of copper and zinc alloys by formation of complex soluble ion.	Cation exchange with hydrogen zeolite. Chlorination. Deaeration.
Dissolved Solids	Dissolved Solids is measure of total amount of dissolved matter, determined by evaporation. High concentrations of dissolved solids are objectionable because of process interference and as a cause of foaming in boilers.	Various softening processes, such as lime softening and cation exchange by hydrogen zeolite, will reduce dissolved solids.
Suspended Solids	Suspended Solids is the measure of undissolved matter, determined gravimetrically. Suspended solids cause deposits in heat exchange equipment, boilers, water lines, etc.	Subsidence. Filtration, usually preceded by coagulation and settling.
Total Solids	Total Solids is the sum of dissolved and suspended solids, determined gravimetrically.	See Dissolved Solids and Suspended Solids.

Reprinted from *Betz Handbook of Industrial Water Conditioning*, Betz Laboratories, Inc., Trevose, PA, 7th ed. (1976).

BIBLIOGRAPHY

Chapter 1

1. Poynton, J. P. "Basics of Reciprocating Metering Pumps." *Chemical Engineering*, May 21, 1979, 156–165.
2. Poynton, J. P. "Metering Pumps: Types and Applications." *Hydrocarbon Processing*, November, 1981, 279–284.

Chapter 3

3. Poynton, J. P. "How to Select Metering Pumps." *Power*, January, 1980, 64–66.
4. Poynton, J. P. "Metering Pumps: Guidelines for Selection to Help Cut Maintenance, Costs." *Pulp & Paper 1981 Buyers Guide*, November 15, 1980, 24–26
5. Poynton, J. P. "Selecting Diaphragm Metering Pumps." *Machine Design*, October 23, 1980, 89–93.

Chapter 4

6. Poynton, J. P. "Design Variables in Chemical Feed Systems, Part I." *Plant Engineering*, April 17, 1980, 141–144.
7. Poynton, J. P. "Design Variables in Chemical Feed Systems, Part II." *Plant Engineering*, May 15, 1980, 137–142.

Chapter 6

8. Brenneisen, S. "Minimize Metering Pump Problems." *Hydrocarbon Processing*, January, 1981, 130–136.

Chapter 7

9. Betz Laboratories. *Handbook of Industrial Water Conditioning*. Betz Laboratories, Inc. Trevose, PA 19047.
10. Poynton, J. P. "Use of Metering Pumps in Power Plants." *Power Engineering*, July, 1981, 76–80.

Chapter 8

11. Calley, A. C., Forster, C. F., and Stafford, D. A. *Treatment of Industrial Effluents*. New York: John Wiley & Sons, Inc., 1973.
12. Cherry, Kenneth E. "Removing Pollutants from Plating Process Wastewaters." *Plant Engineering*, October 15, 1981, 167–170.
13. Poynton, J. P. "How to Introduce Chemicals into Water and Wastewater." *Industrial Water Engineering*, September, 1979, 24–26.

Chapter 9

14. Poynton, J. P. "How to Cut Pumping Costs." *Instruments & Control Systems*, October, 1979, 49–53.

Chapter 10

15. Brenneisen, S. "Minimize Metering Pump Problems." *Hydrocarbon Processing*, January, 1981, 130–136.

INDEX